# What Your Colleagues Are Saying . . .

"Kathryn Chval and her colleagues exquisitely give the readers opportunities to see inside a classroom with multilingual students, develop empathy, and deeply understand effective practice. The book's engaging format provides questions for reflection and strategies to try out, connects research to practice, and compels readers to position students for success."

—Nora G. Ramirez
Executive Director,
TODOS: Mathematics for ALL

"This is the book I've been waiting for! It is powerful. It brings together often-separate critical ideas for teaching multilingual students and weaves them with in-depth explorations of classrooms. We meet remarkable teachers, whose success we can learn from, which will help us reimagine what's possible."

—Lena Licón Khisty
Emerita, University of Illinois Chicago

"A must-read! This book is an excellent resource to closely examine mathematics instruction that affirms multilingual learners' identities, competencies, and growth as learners of mathematics. Far too often multilingual children and their families are positioned in deficit ways that lead to limited learning. This book does the opposite. It seamlessly blends practice and research for a comprehensive look at exemplary mathematics teaching that leverages children's multiple linguistic, cultural, and mathematical strengths. The book offers practical tools and guidance to enhance mathematics instruction, nurture student relationships, and create strong partnerships with families to support and advance multilingual learners in mathematics."

—Julia Aguirre
Faculty Director of Teacher Credential Programs,
School of Education,
University of Washington Tacoma

"This book goes beyond the typical support of the academic language of mathematics for English learners. It provides an in-depth perspective on being more culturally inclusive of English learners and allows educators to reflect on their instructional methodologies in mathematics."

—Alexander L. Tai
Teacher and English Learner Specialist,
Columbia Public Schools

"This book celebrates the brilliance of multilingual learners while also providing evidence-based strategies for teachers. The included cases and activities provide a solid foundation for teachers' growth and exploration into teaching mathematics with multilingual students. This book will help teachers and teacher educators engage in meaningful and humane mathematics instruction with students."

—Zandra de Araujo
Associate Professor, College of Education,
University of Missouri

"This inspiring volume provides resources for mathematics teachers to support mathematics learning for English learners. Using four central principles—assets, empathy, practice, and research—to base the strategies and an impressive array of materials, including student work, the volume illustrates multiple approaches to providing English learners with opportunities to learn important mathematics with understanding."

—Judit Moschkovich
Professor, University of California,
Santa Cruz

"*Teaching Math to Multilingual Students, Grades K–8: Positioning English Learners for Success* takes an asset-based approach toward developing multilingual learners in the classroom. This book clearly demonstrates the nuances of analyzing the mathematical work of multilingual learners while providing examples and strategies for giving useful feedback that is applicable to all learners. Fostering a culture of writing in the mathematics classroom is explicitly taught through a variety of strategies, activities, and teacher practices. Topics such as culturally relevant contexts, crafting language, and family involvement serve to round out this text and provide teachers with a solid resource to support multilingual learners in a layered, thoughtful way."

—Renee Rowan
Second-Grade Teacher,
Skokie, Illinois

"Wondering how to support multilingual learners beyond broad, generic suggestions? This book is it! Through true vignettes, transcripts, pictures, and videos, these authors literally show *how* to support multilingual learners, while engaging you in developing your own capacities to do so. The chapter on positioning learners as leaders is a must-read for every educator! I can't wait to use this book in my work with students and teachers."

—Jennifer Bay-Williams
Author and Professor, University of Louisville

"This book is an excellent resource for opening doors of access to mathematics for multilingual students, particularly those multilingual students who are, in the authors' words, 'silent spectators' of classroom lessons. *Teaching Math to Multilingual Students, Grades K–8* offers strategies and resources that are both research-based and tried, personalized, and polished in real classrooms. The images from those classrooms are compelling, underscoring the importance of an asset-based mentality in teaching multilingual students."

—Mark Driscoll
Coauthor of *Mathematical Thinking and Communication: Access for English Learners*

"This book is a must-have for anyone working with multilingual learners in mathematics. The authors push the reader to reflect through questions and prompts and to take action by trying out the strategies suggested. The authors' deep respect for and asset-based view of multilingual students and their mathematical ideas are evident throughout the whole book. Of particular note is the attention paid to the role of families in the mathematics education of multilingual learners."

—Marta Civil
Professor of Mathematics Education
and Roy F. Graesser Chair–Department of Mathematics,
The University of Arizona

"This groundbreaking book offers practical, research-informed strategies and activities that support all learners. Chval and her colleagues' innovative approach positions multilingual learners as potential classroom leaders in challenging mathematics learning. Even newcomers to English are invited to draw on all their meaning-making resources to participate in the mathematics classroom."

—Mary J. Schleppegrell
Professor of Education,
University of Michigan, Ann Arbor

"This book provides a powerful tool for teachers as they engage multilanguage learners in language and mathematics. Each chapter is a wonderful compilation of research and practice that unpacks the strategies that will empower teachers to build upon the unique strengths and knowledge that multilingual students bring to the classroom."

—Amy Stephens
Senior Program Officer,
Board on Science Education,
The National Academies of Sciences,
Engineering, and Medicine

# Teaching Math to Multilingual Students
## The Book at a Glance

Multilingual learners deserve the same social and academic opportunities to learn and be successful as their English-speaking peers. This book takes an asset-based, empathetic, practical, and research-based approach to help you position multilingual learners as leaders in your mathematics classes so that they may strive for success. You will be aided in your journey through:

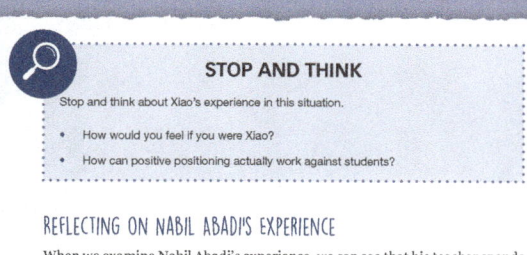

**Reflection Questions** to help you examine your teaching practice.

---

**STOP AND THINK**

Stop and think about Xiao's experience in this situation.

- How would you feel if you were Xiao?
- How can positive positioning actually work against students?

### REFLECTING ON NABIL ABADI'S EXPERIENCE

When we examine Nabil Abadi's experience, we can see that his teacher spends time observing Nabil's interactions first. She does not make assumptions about what might be causing Nabil's frustration, and she does not allow a negative storyline such as "troublemaker" to be instituted. Instead, she observes and notices how particular students dominate small-group talk and activities. Recognizing that this is harmful for every student, she uses the power of her position as the teacher to make space for Nabil to speak. By doing so, she positions Nabil as someone with important contributions to make and one who has ideas that others should listen to and consider. This also provides opportunities for her to learn about Nabil's mathematical sensemaking and give Nabil opportunities to use English to share his mathematical ideas.

---

**Reflect**

- Imagine you were to record and study your teaching. What would you find about your positioning of multilingual learners?

- What strategies will you use more often to position multilingual learners as leaders in your classroom?

- What storylines are present in your mathematics classroom? Are there storylines you want to alter? If so, how will you alter them or promote new ones?

---

**Instructional Strategies** to assess student understanding, partner students, appropriately challenge them, identify and draw out their strengths, and encourage and motivate student participation.

### STRATEGIES FOR PROMOTING CLASSROOM DISCOURSE

Mercer (1995, p. 32) identified the following strategies that promote and initiate classroom discourse:

▶ Make a declarative (open-ended or provocative) statement that invites a rejoinder or disagreement;

▶ Invite elaboration ("Could you say a bit more about that?");

▶ Admit perplexity when it occurs, whether about the topic itself or about a pupil's contribution to it;

▶ Encourage questions from pupils (rare in many classrooms); and

▶ Maintain silence at strategic points (Dillon [(1982), another classroom researcher,] suggests that three to five seconds may be enough to draw in another pupil's contribution or encourage the previous speaker to elaborate on what was said).

**Try It! 4.2**

Choose Partners for Multilingual Learners

Pat and Sam are two new students who will join your class tomorrow. Here is some information about each of them.

Pat is a third-grade student who has just joined your class after recently moving to Missouri from New Mexico, where Pat's family lived for two years. Pat's family relocated to be closer to extended family and possible new financial opportunities as the family's financial situation was bleak in New Mexico. Pat is the middle child of five, who all live with their mother. Since moving to Missouri, Pat has not adjusted well socially and has been unable to make friends. Pat is very quiet in class, appears timid, and does not participate in class discussions.

Sam is also a third-grade student who has just joined your class after moving to Missouri from Mexico. Sam's family relocated for new job opportunities. In Mexico, Sam's family was financially stable and had the privilege to travel annually. Sam's family is composed of a younger sister, Nancy, and both parents. Since moving to Missouri, Sam has had trouble integrating into the school community and making friends. In class, Sam has not performed well academically and has picked a fight with another student.

- What qualities would you look for when selecting partners for Pat and Sam?

- Why are these qualities important for each of these specific students?

**Try It! Activities** that invite you to apply the strategies.

**Vignettes and Transcripts** of real conversations between teachers and students as well as between teachers and the authors themselves.

**Transcript 2.3**

| TRANSCRIPT | | POSITIONING ACTS |
|---|---|---|
| Ms. Bristow: | You know, I saw some kids who did a much better job than I did drawing efficient pictures. So, I wanted to talk to you—I wanted a few of those kids to come up. Lorena, you're my first friend to come share. We're going to talk about number two. Ms. Bristow gave six pieces of candy to Jake, Avery, Carl, and Erica. How much candy did she give out all together? Tell us about your picture. | Positioned Lorena as an efficient drawer in front of her peers<br><br>Positioned Lorena as a student who can explain her thinking to peers |
| | *Lorena's work is shown on the board. She has the following drawn on her paper:* | Scanned Lorena's work so she could use gestures to enhance her explanation of her strategy |
| | $6 + 6 + 6 + 6 = 24$ | |
| Lorena: | Well, first I made four groups that have six . . . I did 6 plus 6 plus 6 plus 6 equals 12, I mean 24. And then I added. I had to draw a picture of six and then I added them and . . . | Did not interrupt Lorena as she self-corrected when describing her strategy |
| Ms. Bristow: | So, um, your picture—did it take very long for you to draw that picture? | Asked Lorena to reflect on the efficiency of her picture in front of the class |
| Lorena: | [*Shakes head no*] | |
| Ms. Bristow: | No. And you were able to quickly count that there were 24 of them? Wonderful. That's very efficient. Do you guys have comments or compliments for Lorena? | Repositioned Lorena as an efficient drawer in front of her peers; expected peers to attend to Lorena's mathematical thinking |

*Source:* Smith (2018).

# Teaching Math to Multilingual Students

## Grades K–8

This book is dedicated to Sara Martínez, an exceptional teacher who was the impetus and inspiration for the work.

We also dedicate this book to multilingual learners and their families across the United States and the world because they enrich our schools, communities, and cultures with their diverse perspectives, experiences, and knowledge. We are grateful to the families who graciously opened their homes so others could learn from them.

# Teaching Math to Multilingual Students

## Positioning English Learners for Success

### Grades K–8

Kathryn B. Chval,
Erin Smith,
Lina Trigos-Carrillo, and
Rachel J. Pinnow

A Joint Publication

*For information:*

Corwin
A SAGE Company
2455 Teller Road
Thousand Oaks, California 91320
(800) 233-9936
www.corwin.com

SAGE Publications Ltd.
1 Oliver's Yard
55 City Road
London, EC1Y 1SP
United Kingdom

SAGE Publications India Pvt. Ltd.
B 1/I 1 Mohan Cooperative
    Industrial Area
Mathura Road, New Delhi 110 044
India

SAGE Publications Asia-Pacific Pte. Ltd.
18 Cross Street #10-10/11/12
China Square Central
Singapore 048423

Publisher, Corwin Mathematics: Erin Null
Associate Content
    Development Editor: Jessica Vidal
Content Development Editor: Desirée Bartlett
Editorial Assistant: Caroline Timmings
Production Editor: Tori Mirsadjadi
Copy Editor: Melinda Masson
Typesetter: Integra
Proofreader: Liann Lech
Indexer: Integra
Cover Designer: Scott Van Atta
Marketing Manager: Maura Sullivan

*Library of Congress Cataloging-in-Publication Data*

Names: Chval, Kathryn B. (Kathryn Bouchard), author.
Title: Teaching math to multilingual students, grades K-8 : positioning English learners for success / Kathryn B. Chval, Erin Smith, Lina Trigos Carrillo, and Rachel J. Pinnow.
Description: Thousand Oaks, California : Corwin, 2021. | Includes bibliographical references and index.
Identifiers: LCCN 2020041785 | ISBN 9781071810842 (paperback) | ISBN 9781071810835 (ebook) | ISBN 9781071810828 (ebook) | ISBN 9781071810811 (adobe pdf)
Subjects: LCSH: Mathematics—Study and teaching (Elementary) | Mathematics—Study and teaching (Middle school) | Multilingualism. | Language and education. | Language arts—Study and teaching—Correlation with content subjects.
Classification: LCC QA135.6 .C534 2021 | DDC 372.7/044—dc23
LC record available at https://lccn.loc.gov/2020041785

24 25 26 27 10 9 8 7 6 5 4

# Contents

## 5. Engage Multilingual Learners Through Culturally Relevant Contexts

## 6. Reach Multilingual Learners With Visuals and Gestures

## 7. Analyze Mathematical Work of Multilingual Learners

## 8. Investigate Meanings to Enhance Multilingual Learners' Language Development

## 9. Use Your Discourse Strategically to Enhance Multilingual Learners' Opportunities to Learn

## 10. Foster a Culture of Writing in the Mathematics Classroom

## 11. Develop Writing in Mathematics for Multilingual Learners

## 12. Enhance Curriculum Materials for Multilingual Learners

## 13. Engage With Parents and Families of Multilingual Learners

Note From the Publisher: The authors have provided video and web content throughout the book that is available to you through QR (quick response) codes. To read a QR code, you must have a smartphone or tablet with a camera. We recommend that you download a QR code reader app that is made specifically for your phone or tablet brand.

# Preface

Throughout our careers, we have had the privilege of collaborating with and researching inspirational teachers. Ms. Sara Martínez is one of those exceptional teachers. She opened her elementary mathematics classroom to us so that we could learn from her practice. We are grateful that the results from that first study in her classroom became the seed for robust and innovative research that was replicated in other classrooms in different grade levels and contexts.

Ms. Martínez is a teacher who establishes conditions for student success, where every child is respected and challenged, has the flexibility to solve mathematics problems in several ways, and is given the opportunity to communicate mathematical thinking in her classroom. Everyone who enters Ms. Martínez's classroom is mesmerized by what her fifth graders can do mathematically. Ms. Martínez creates a classroom community whose culture is characterized by respectful challenge, agreement and disagreement, and argument. The students listen closely to each other's ideas, build on each other's work, and can complete a peer's problem-solving strategy at a moment's notice. In this book, we share examples of Ms. Martínez's practice during her 20th year of teaching in Chicago. We recorded lessons with her class of 24 fifth graders during a year when her school reported 96.8% of the students as low-income; 96.9% as Hispanic; 46% with limited English proficiency (the term the state uses; however, not a term we endorse); and a mobility rate of 21.5%.

During our professional development sessions with teachers around the country, we have shared lessons learned from inspirational teachers such as Ms. Martínez. As a result, we consistently receive requests for specific strategies and materials that facilitate the engagement of children learning mathematics in languages that differ from their native languages—in other words, multilingual learners. This book was born out of that demand.

Multilingual learners deserve the same social and academic opportunities to learn and be successful as their English-speaking peers. All students should learn how to interpret the meaning of problems, make conjectures, analyze mathematical thinking and solutions, monitor and evaluate their progress, and understand the approaches of others in comparison with their own. These expectations emphasize the vital role of language and communication in solving mathematical problems, including the different domains of language (i.e., reading, writing, speaking, and listening) in developing mathematical thinking, and demonstrating knowledge in classroom interactions. As teachers, we must facilitate access, participation, and success for multilingual learners. To do this effectively, we must recognize that multilingual learners require opportunities to learn content while simultaneously developing a new language.

# OUR UNIQUE PERSPECTIVE

This book has four underlying principles. This book is . . .

▶ **Asset-based:** Multilingual learners "bring new perspectives and resources to the classroom through their participation and sharing of experience that can benefit their peers" (National Academies of Sciences, Engineering, and Medicine, 2018a, p. 21). Multilingual learners are intellectual leaders of classrooms. Everyone can learn from their complex knowledge and experience when we position students' language and culture as valuable resources for learning (Ladson-Billings, 2014; Orellana, 2016). Therefore, we encourage you to draw on students' academic success, social and cultural identities, and family participation (i.e., assets or strengths) to make your lessons more interesting for all of the students in your classroom (Kobett & Karp, 2020; Paris, 2012).

▶ **Empathy-based:** We focus on developing empathy by asking you to imagine yourself in the shoes of multilingual learners, parents/families of multilingual learners, or teachers of multilingual learners. We set up situations and provide questions that facilitate reflection and the consideration of different perspectives. For example, you will consider times when you were reluctant to speak in public or share your thinking, and when you were in situations where information was confusing due to the cultural context such as viewing a cricket match.

▶ **Practice-based:** Each chapter includes content from our studies of teaching practice in elementary classrooms. During our research, we recorded and transcribed mathematics lessons, interviews and lesson planning sessions with teachers, and interviews with multilingual learners and their parents. We also collected copies of student work. We integrate different examples of teaching practices and strategies designed to illustrate diverse, effective ways of teaching. You will find examples of teachers with different levels of teaching experience, ranging from 1 year to 20 years, from a variety of school settings, including rural and urban contexts.

▶ **Research-based:** Too often, we hear deficit perspectives and narratives about teaching and teachers—in other words, what teachers do not know or cannot do. As we researched classroom teaching, we identified what the participating teachers wanted to learn about enhancing their teaching practice and then facilitated that journey along with them. We drew from research on how people learn (National Academies of Sciences, Engineering, and Medicine, 2018b) to build on what teachers already knew to develop the specialized knowledge and competencies to teach mathematics to multilingual learners. Along the way, we know we learned more from the teachers, parents, and multilingual learners than we could have imagined. The insights gleaned from this research make this book unique.

## AUDIENCE

A range of audiences will benefit from the content of this book, including novice and experienced K–8 teachers; mathematics coordinators, coaches, and supervisors; curriculum coordinators; mathematics teacher educators; professional development facilitators; and faculty in teacher preparation programs. We encourage you to facilitate conversations about the ideas in this book across your schools, organizations, and communities. A key factor in teaching multilingual learners is to enhance the cultural environment of schools and to foster the critical conversations necessary to build strong schools. In the development of this book, we discussed the chapters with preservice teachers, English as a second language (ESL) and mathematics coordinators, practicing teachers, and administrators over the course of several years. Their contributions and insights have helped create a book that addresses the needs of a wide range of stakeholders.

## ORGANIZATION

Working with multilingual learners is complex. In this book, you will not find quick fixes. Instead, each chapter presents a different aspect of teaching mathematics to multilingual learners that must be thoroughly considered. The strategies in the book will help you draw out the strengths and knowledge of multilingual learners and other students. These practices will profoundly affect every student in the classroom so that when they are adults, they too will function with an asset-based mentality when meeting people who differ from them or who are marginalized by society.

Within each chapter, you will encounter

▶ **reflection questions** to help you examine your teaching practice;

▶ **strategies** to assess student understanding, appropriately challenge students, partner students, identify and draw out students' strengths, and encourage and motivate student participation;

▶ **Try It! activities** that invite you to apply the strategies; and

▶ excerpts from **transcripts** of conversations between teachers and students as well as between teachers and ourselves.

After reading this book, you will be able to

✓ support the development of mathematics and language for multilingual learners;

✓ enhance curriculum materials to ensure they are challenging and accessible for multilingual learners; and

✓ position multilingual learners for success as individuals, in groups, and in whole-class settings.

We encourage you to think about the following questions as you read each chapter:

▶ *How am I ensuring the academic success of multilingual learners?*

▶ *In what ways am I valuing, sustaining, and learning from multilingual learners and their families' heritage, knowledge, and culture?*

▶ How am I working with multilingual families so that they are partners in educating their children?

# Acknowledgments

This book was a labor of love. It would not have been possible without the commitment, generosity, and dedication of our collaborators, funders, partners, and family. We thank each of you for your time, talent, trust, and intellect.

This book started when the National Science Foundation (NSF) funded the project CAREER: A Study of Strategies and Social Processes That Facilitate the Participation of Latino English Language Learners in Elementary Mathematics Classroom Communities on July 15, 2009, under the direction of Dr. Ferdinand D. Rivera. This material is based on work supported by the NSF under Grant Number DRL-0844556. Any opinions, findings, and conclusions or recommendations expressed in this material are those of the authors and do not necessarily reflect the views of the NSF.

We are grateful for mentors who influenced the design of the project, including Óscar Chávez, PhD; Guillermo Solano-Flores, PhD; Maria Araceli Ruiz-Primo, PhD; Marcela Chávez; Marta Civil, PhD; Mark Driscoll, PhD; Lena Licón Khisty, PhD; John Lannin, PhD; Fran Arbaugh, PhD; and Norma López-Reyna, PhD. A special thank-you to Dr. Óscar Chávez, who was instrumental in creating the student-worn cameras and codesigning the pilot study in 2005.

We are grateful for the research team who labored to collect and analyze the data from 2009 to 2020, including Kathryn B. Chval, PhD; Liza Cummings, PhD; Anne Estapa, PhD; Sarah Hicks, PhD; Maryann Huey, PhD; Oscar Rojas Perez, PhD; Rachel J. Pinnow, PhD; Amaya Praschan; Erin Smith, PhD; Rukiye Didem Taylan, PhD; Amanda Thomas, PhD; Lina Trigos-Carrillo, PhD; and Luz Edith Valoyes-Chávez, PhD.

We are grateful for our school collaborators, including district and school administrators (i.e., superintendents, principals, ESL coordinators, curriculum coordinators, assessment directors, translators, and office professionals who greeted us during each visit). A special thank-you to the teachers who spent countless hours with us. They trusted us with their most valued space and gave us permission to listen to their planning, teaching, and thinking. We will be forever grateful for their courage. We would also like to thank the teachers, ESL coordinators, and preservice teachers who used the materials in professional development and course settings so that we could improve the content and relevancy for teachers with different levels of experience in different contexts. Maggie Hackett and Tabetha Finchum provided a careful review of the materials from the perspectives of teachers and professional development facilitators. As you can see, the development of this book would not have been possible without this incredible team.

Finally, we are grateful for our family and friends who love and encourage us, especially when we escape to write.

# Publisher's Acknowledgments

Corwin would like to thank the following individuals for taking the time to provide their editorial insight and guidance:

Melissa Black
Associate Dean
Progressive Education Institute
New York, NY

Diana Ceja
Former Board President, TODOS: Mathematics for ALL
Administrator, Riverside County Office of Education
Riverside, CA

Cathy Martin
Associate Chief of Academics
Denver Public Schools
Denver, CO

Rosamaria Murillo
Principal
Ladera Palma Elementary
La Habra, CA

Jennifer Newell
Mathematics Compliance Specialist
Istation
Dallas, TX

Venessa Powell
Mathematics Assessment and Test Development Officer
The Student Assessment Unit
Kingston, Jamaica

Donella Sherry
Math Consultant
Maryville, MO

# About the Authors

**Dr. Kathryn B. Chval** is a professor of mathematics education at the University of Missouri. Kathryn's commitment to educational solutions in mathematics education is rooted in her early experiences teaching elementary grades in underresourced schools in the United States. Kathryn's research focuses on effective preparation models and support structures for mathematics teachers, effective elementary mathematics teaching for multilingual learners, and curriculum standards and policies. Prior to joining the University of Missouri in 2003, Kathryn was the acting section head for the Teacher Professional Continuum Program in the Division of Elementary, Secondary, and Informal Education at the National Science Foundation (NSF). She worked at the University of Illinois at Chicago from 1989 to 2001 as the codirector on mathematics curriculum development projects and systemic change projects funded by the NSF. She has served as an investigator on projects including the Center for the Study of Mathematics Curriculum, ALL Learn Mathematics, and Collaborative Research: Parents, Teachers, and Multilingual Children Collaborating on Mathematics Together. Additionally, she is the recipient of the prestigious NSF Early Career Award for a project titled A Study of Strategies and Social Processes That Facilitate the Participation of Latino English Language Learners in Elementary Mathematics Classroom Communities. Kathryn's leadership, research, and service have been recognized with several awards, including the Association of Mathematics Teacher Educators (AMTE) Early Career Award, the INSIGHT Into Diversity Giving Back Award for administrators, the University of Illinois at Chicago College of Education Alumni Award, and the NSF Director's Award for Program Management Excellence.

**Dr. Erin Smith** is an assistant professor of mathematics education in the School of Education at the University of Southern Mississippi. Erin received her PhD in learning, teaching, and curriculum from the University of Missouri. Her interest and passion for increasing access to high-quality mathematics for multilingual learners stems from her work teaching mathematics and English. Erin's research examines the practices of exemplary monolingual teachers of multilingual learners, the preparation of preservice teachers for diverse learners, and the facilitation of parent/guardian collaborations designed to advance children's success in school mathematics. Prior to joining the University of Southern Mississippi, Erin was an instructor of mathematics at Zayed University in Abu Dhabi, United Arab Emirates, and a teacher of English as a foreign language in Seoul, South Korea. Erin is a member of the National Council of Teachers of Mathematics, TODOS: Mathematics for ALL, and the American Educational Research Association. She is a McNair Scholar; North Carolina State University Building Future Faculty Fellow;

Association of Mathematics Teacher Educators (AMTE) Service, Teaching, and Research (STaR) Fellow; and Service-Learning Fellow at the University of Southern Mississippi.

**Dr. Lina Trigos-Carrillo** is the chair of the Department of Psychology of Development and Education and an associate professor in the School of Psychology at Universidad de La Sabana in Chía, Colombia. Lina is a professor of literacy education, community education, and qualitative research methods. She received a PhD in learning, teaching, and curriculum from the University of Missouri, and she participated as a research postdoctoral fellow in the project Strengthening Equity and Effectiveness for Teachers of English Learners (SEE-TEL), a National Professional Development Program grant from the U.S. Department of Education. Lina's research focuses on critical sociocultural perspectives to writing and community/family literacies of people of color and multilingual learners across the Americas. She has conducted qualitative research with multilingual families in the United States and with diverse families and communities in Mexico, Costa Rica, and Colombia. She designs culturally sustaining professional development based on her research experiences in global contexts.

**Dr. Rachel J. Pinnow** is an associate professor at the University of Missouri in the College of Education. She currently serves as the emphasis area leader of the Teaching English to Speakers of Other Languages (TESOL) program in the Department of Learning, Teaching, and Curriculum. Rachel received her PhD in language and literacy education with a specialization in teaching additional languages (TAL) from the University of Georgia. Rachel grew up speaking multiple languages; taught English in Dalian, People's Republic of China, for several years; and continues to work with multilingual learners in various educational contexts such as South Africa and many U.S. communities. As a multilingual speaker and an applied linguist, Rachel focuses her research on second language acquisition, classroom interaction, multimodal communication and analysis, social semiotic theory, and positioning theory. Her work seeks to provide insight to how classroom interactions can provide affordances for content and second language learning for multilingual learners. She currently serves as principal investigator on a National Science Foundation–funded grant, Collaborative Research: Parents, Teachers, and Multilingual Children Collaborating on Mathematics Together, that examines positioning among multilingual learners, their family members, and teachers during mathematics education and instruction.

# CHAPTER 1
## OUR HOPE FOR MULTILINGUAL LEARNERS

## TEACHERS WHO INSPIRE

Most of us teachers can recall someone who influenced our entry into education, taught us how to become better teachers, and inspired us to enhance the lives of others. We authors would love to hear the stories about the individuals who inspired you because those are the stories that give us hope. Your decision to read this book tells us you are looking for approaches to enhance the participation and success of multilingual learners. Thank you for making that investment!

Early in my career, I (Kathryn) worked with a variety of elementary teachers as I was writing mathematics curriculum, teaching in K–8 schools, and facilitating professional learning sessions. One of the most instrumental relationships was with a fifth-grade teacher, Ms. Sara Martínez. Ms. Martínez is a bilingual teacher, fluent in English and in Spanish, who is excellent at connecting with both children and their families. She was known as the teacher who held very high expectations for her students. I collaborated with Ms. Martínez to conduct a research study, which involved observing her classroom 60 times during the school year: five times in Week 1, three times a week in Weeks 2 to 6, and one to two times per week for the rest of the school year. I documented a careful record of what happened in the classroom by compiling field notes as well as collecting student work each week in the curriculum materials and samples of writing assignments. A total of 119 mathematics lessons were audio recorded. Ms. Martínez's classroom was composed of students whose primary language was Spanish. At the time the study took place, Ms. Martínez had a self-contained class of 24 students who represented a wide range of proficiencies in Spanish and English. As with most fifth-grade language learners, the students were still developing in academic English (as opposed to conversational English) proficiency. Figure 1.1 includes the median grade equivalent for the Iowa Test of Basic Skills (ITBS) reading test before entering Ms. Martínez's classroom compared to the other fifth graders in the school, district, and national norm.

**Figure 1.1**   Median Grade Equivalent (Reading) Prior to Entering Ms. Martínez's Classroom

| COMPARISON GROUPS | END OF GRADE 4 |
| --- | --- |
| Ms. Martínez's Class | 3.7 |
| Other Fifth Graders in the School | 4.0 |
| Fifth Graders in the District | 4.2 |
| National Norm | 4.8 |

**Figure 1.2**  Growth in One Year Measured by Median Grade Equivalent (Math Total)

| COMPARISON GROUPS | END OF GRADE 4 | END OF GRADE 5 | GAIN |
|---|---|---|---|
| Ms. Martínez's Class | 4.3 | 6.1 | 1.8 |
| Other Fifth Graders in the School | 4.6 | 5.8 | 1.2 |
| Fifth Graders in the District | 4.6 | 5.6 | 1.0 |
| National Norm | 4.8 | 5.8 | 1.0 |

*Source:* Razfar, Khisty, and Chval (2011).

Figure 1.2 demonstrates the mathematical gains that Ms. Martínez's students made in her classroom as measured by the ITBS mathematics assessment.

As we can see from the fourth-grade column, the average child in Ms. Martínez's classroom was half a year behind the expected 4.8. Five of the 24 students (20.8%) performed at the 4.8 level or above. After just eight months in Ms. Martínez's classroom (fifth-grade column), her students outperformed the other two groups, and 15 of the 24 students (62.5%) performed at the 5.8 level or above. Overall, Ms. Martínez's students accomplished a great deal in a short amount of time as evidenced by their performance on not only the ITBS, but also other measures collected during the study.

Each day I observed Ms. Martínez's classroom was memorable, but the first day of the school year particularly stands out. One common misconception about teaching multilingual learners is that they should not be asked to engage in challenging academic work, such as complex mathematics problem solving, until they are at grade level in English language proficiency. Due to this misconception, multilingual learners often languish in academic content classrooms where they continue to fall behind in both academic content learning and second language acquisition as the years of study necessary for grade-level proficiency pass. Since multilingual learners are learning how academic language and academic content work by using these resources to engage in classroom tasks and activities, waiting until they reach grade-level proficiency in English can actually keep them from making progress. The social ramifications are alarming as these learners can come to believe that (because they are not being challenged with grade-level work) they must be deficient in some way that disqualifies them for the same success as their peers.

Fortunately for her students, Ms. Martínez created a challenging and supportive environment for all students where multilingual learners flourished. The emphasis in Ms. Martínez's classroom was always on solving challenging problems, explaining how to solve hard problems, identifying more efficient ways of solving problems, and investigating more interesting ways of solving problems. When students progressed beyond Ms. Martínez's own mathematical knowledge, which was very strong, she was not deterred. In order to shape her classroom as a place where learning was valued, she was very open about modeling how learners, herself included, admit when they do not know the answer or how to proceed in solving a problem. Ms. Martínez made a practice of tackling advanced mathematics problems that she herself struggled with in order to show that "the reason we are in school is we are learning. If we make a mistake, that is great. Let's put it up on the board so that we can figure out how to fix it." This approach also made the *process* of learning more important than producing the product of a correct mathematical answer. *How* you got to the answer mattered, including where you might have gone awry or faced difficulties in problem solving.

Ms. Martínez also operated in ways that were countercultural in that problem solving was not viewed through an individualized lens—something that occurred inside the learner alone and belonged to the learner alone. Rather, learning was socially distributed, a classroom community process that required that everyone participate, share their knowledge and questions, and share their struggles. By creating a community of learners, students achieved more, not less, as the test scores of her students after one year indicate.

When I showed a video of Ms. Martínez's teaching at a professional conference, it was clear no one had seen anything like it. I began to wonder: How can I provide opportunities for other teachers to learn about Ms. Martínez's practice? I will always be grateful for her willingness to open her classroom so that I could share her strategies teaching mathematics to multilingual learners through transcripts of her teaching. To read more about Ms. Martínez's teaching practice, see Chval (2004, 2012); Chval and Chávez (2011); Chval and Khisty (2009); Khisty and Chval (2002); Morales, Khisty, and Chval (2003); and Razfar, Khisty, and Chval (2011).

## STUDYING TEACHER PRACTICE

I (Kathryn) designed additional research studies, funded by the National Science Foundation, that involved a variety of elementary teachers so other educators could learn from Ms. Martínez. The research studies that followed involved children wearing video cameras mounted on hats with Velcro® and then more sophisticated wireless video cameras that captured mathematics teaching and learning as shown in Figure 1.3 and Figure 1.4. In the early stages in 2005, Óscar Chávez and I tried out this approach in a first-grade classroom and a fourth-grade classroom to determine if the video cameras would capture useful data.

Then I designed a research study where I collaborated with four third-grade teachers during three academic years from 2009 to 2012. Each week, I met with the teachers to introduce ideas that would influence the design and enactment of instruction, in relation to multilingual learners (i.e., planning sessions). After the planning sessions, the research team (including Rachel Pinnow and Lina Trigos-Carrillo) videotaped two mathematics lessons in each classroom. Then I met with each teacher to debrief lessons and discuss video clips that were filmed during the past week (i.e., debrief sessions). Each session was shaped as a conversation about teaching and learning, rather than a

**Figure 1.3** Third-Grade Girl Wearing a Video Camera Mounted on Her Hat

**Figure 1.4** Third-Grade Boy Wearing a Wireless Video Camera Demonstrating His Approach With a Task

*Source:* Pinnow, R. J., & Chval, K. B. (2015). *Linguistics and Education*. Columbia, MO: Elsevier. Used with permission.

*Source:* Chval, K. B., Pinnow, R. J., & Thomas, A. (2015). *Mathematics Education Research Journal*. Used with permission.

directive of how the teachers should teach. After each planning session, the teachers would reflect on the conversation, design their own lessons, and create lesson materials. The professional development involved in this process included all the components that Garet and colleagues (2001) identified as critical for effective professional development. For example,

- I worked with the teachers consistently for three years;

- I focused on the content I wanted the teachers to learn—teaching mathematics to multilingual learners in elementary classrooms;

- I integrated lesson planning to connect the work with teachers' daily experiences and constraints; and

- I integrated discussions and lesson planning so I could assess teachers' prior knowledge and experiences as I thought about what kinds of questions to pose and in what ways I could facilitate their thinking about teaching multilingual learners, especially through the selection of videos filmed by students in their classrooms.

One third-grade teacher, Laura McKinney, reflected on how the use of the student cameras helped her grow during the first year:

> " *The first time I watched a video filmed with the head cameras, I was shocked. I couldn't believe the things I missed even though I was right there! It concerned me at first, but as the year went on, I realized some great things were happening. I was able to see student interaction without the students feeling the need to please me, because I wasn't hovering over them. Another benefit [of the video cameras] is the opportunity to see student weaknesses. When students take 10 minutes to start an activity, I know they are struggling*

*somewhere. I can also see what exactly the students [multilingual learners] are doing in the process of working on a problem. I can see their mistakes as they make them and am better able to understand why they make those mistakes.* **99**

The data generated from the student-worn cameras and discussions with teachers provided insight to teachers' and multilingual learners' experiences.

Throughout this book, you will read transcripts from these interactions as well as from the mathematics lessons that they taught. We use pseudonyms when we reference teachers and students in the book. As we worked with teachers and analyzed the data, we noticed that children in classrooms of teachers who teach like Ms. Martínez also learn to value every person, all the languages they speak, and what they contribute to the classroom community. We are so grateful for the teachers who were willing to invest time to learn how to more effectively teach mathematics to multilingual learners; open their classrooms to multiple video cameras; and, most importantly, share their practice with other teachers. See Figure 1.5 for more information on the classrooms studied.

**Figure 1.5**   Teachers Involved in the Studies

| TEACHER | YEARS OF EXPERIENCE | GRADE | SCHOOL | # OF STUDENTS | # OF MLLS | LANGUAGES |
|---|---|---|---|---|---|---|
| Sara Martínez | 20 | 5 | Large urban district 96.8% low-income 96.9% Hispanic 46% limited English proficient 21.5% mobility rate | 24 | 24 | Spanish |
| Courtney Bristow | 2 | 3 | Small city 6.6% Hispanic (year 1) 9.5% Hispanic (year 2) 58.1% free and reduced-price lunch | 22 | 3 | Spanish |
| Kari Reams | 2 | 1 | Small city 3% Hispanic 8% Asian 16.9% free and reduced-price lunch | 22 | 7 | Spanish Korean Chinese |
| Roger Jones | 15 | 4 | Small city 3% Hispanic 8% Asian 16.9% free and reduced-price lunch | 18 | 1 | Spanish |

| TEACHER | YEARS OF EXPERIENCE | GRADE | SCHOOL | # OF STUDENTS | # OF MLLS | LANGUAGES |
|---|---|---|---|---|---|---|
| Laura McKinney | 1 | 3 | Rural industrial<br>22% Hispanic<br>76% free and reduced-price lunch | 20 | 4 | Spanish |
| Jessica Barnes | 3 | 3 | Rural industrial<br>22% Hispanic<br>76% free and reduced-price lunch | 21 | 3 | Spanish<br>Russian |
| Cindy Keller | 18 | 3 | Rural industrial<br>22% Hispanic<br>76% free and reduced-price lunch | 21 | 9 | Spanish<br>Russian |

*Note:* MLLs = multilingual learners.

## INTERACTING WITH MULTILINGUAL LEARNERS AND THEIR FAMILIES

During the three-year study involving Ms. Bristow, Ms. Keller, Ms. McKinney, and Ms. Barnes, we also interviewed the multilingual learners and their parents. We were well aware of misconceptions about children whose first language is not English among preservice teachers (see Chval & Pinnow [2010] and Vomvoridi-Ivanovic & Chval [2014] for examples). In addition, Ogbu and Simons's (1998) argument that "the treatment of the minorities in the wider society is reflected in their treatment in education" (p. 161) suggested that we needed to listen to the stories of the families involved. In an effort to provide counter-stories to the deficit-oriented comments we had heard from some educators, we interviewed parents to learn more about the families of multilingual learners. We heard the incredible love that the families had for their children and that they would give up everything—would leave their homeland to come to the United States—so that their children could have what they hoped would be a better life. During these conversations with families, many truths were made evident. As noted, we are aware that misconceptions exist regarding multilingual families. To dispel some of those misconceptions, we created Figure 1.6. We are eager to hear from you what you would add to the list.

**Figure 1.6**  Misconceptions About Multilingual Families

| MISCONCEPTION | REALITY |
|---|---|
| Families do not care to attend meetings at schools. | Parents may work multiple jobs to provide for their families in the United States and back home. They may not be able to miss work during the school day or evenings to attend school meetings. |
| Families are not interested in information disseminated by the school. | Multilingual families may have limited literacy levels in English and may not understand the flyers, emails, and homework assignments. |
| Families are not literate. | They may be extremely well educated, but in a different language. |
| Families do not value education. | Multilingual families value education so much they are willing to leave their homeland and (sometimes) live in poverty in the United States so that their children can enjoy quality and safe educational experiences. |

Parents make very difficult choices for important reasons that are often unknown to outsiders. You will also have opportunities to learn from parents as you engage in the content of this book. Thank you for participating in these important conversations with us. We know they will influence your teaching of mathematics, the participation of multilingual learners in your classroom, and your engagement with families.

# CHAPTER 2
## POSITION MULTILINGUAL LEARNERS AS CLASSROOM LEADERS

### Key Concepts

In this chapter, you will

- ✓ learn about the idea of positioning, its importance, and how positions can affect multilingual learners in mathematics classrooms.

- ✓ identify how teachers position students in the classroom.

- ✓ identify instructional strategies for enhancing multilingual learners' positions in the mathematics classroom.

As we've worked in classrooms, we have noticed situations where multilingual learners are positioned by others (i.e., students and teachers) in ways that do not support their mathematics learning. Yet, in other cases we've seen teachers who have positioned multilingual learners as competent mathematical thinkers, role models, and teachers. To illustrate these differences, consider the following examples.

## VIGNETTE: XIAO LI

Xiao Li is a multilingual learner from mainland China who just joined a fourth-grade classroom in the United States. Li is Xiao's family name. Xiao means "small or little," so Xiao is probably the smallest or youngest member of the Li family. Xiao has been taught in her home country not to speak unless the teacher calls on her, so she is a strong listener who does not raise her hand to contribute since she is waiting on the teacher to select her to answer questions.

Xiao's teacher believes Xiao is "just shy" and will "come out of her shell" as she grows more comfortable in her new environment, so the teacher chooses not to call on Xiao in order to give her time to adapt to U.S. classroom norms. In this vein, Xiao's teacher uses small groups in order to provide students with opportunities to practice solving mathematics problems with others, which the teacher also hopes will provide an environment in which Xiao feels more comfortable talking with her peers. On Xiao's second day in this classroom, her teacher organizes students into small groups, saying with a smile and a chuckle, "Xiao, you will work with Nicole, Evan, and Tommy as they are strong mathematicians and I know you'll need to be challenged!"

## VIGNETTE: NABIL ABADI

Nabil Abadi is a Syrian refugee who has been attending school in the United States for three months. The name Nabil means "noble." He is quiet, but sometimes erupts in frustration when working with his peers. Nabil's teacher begins to closely observe Nabil as he works with his peers to identify the source of the problem. She finds that Nabil's peers do not share the mathematical tools with him, talk over Nabil when he tries to provide answers, and even take Nabil's worksheets from him, writing on or erasing his work.

Nabil's teacher also notices that when she joins Nabil's small group, other more vocal or dominant students tend to nominate themselves as leaders of the small group and take over providing the explanations for the group's mathematical problem solving. Nabil is rarely able to gain the floor in order to speak or explain his ideas in these instances. Nabil's teacher begins to take a different approach. She initiates her interaction with the small group by selecting Nabil as the leader of the group who should explain the group's mathematical answers. When Nabil's peers interrupt him or try to talk over him during his explanation, his teacher stops them, looks directly at Nabil, and says, "Keep telling us your ideas." Over time, Nabil's peers stop interrupting him when he talks or when there are gaps in his speech as he thinks through his answers in English. Nabil's teacher also begins to gain a better idea of Nabil's mathematical thinking, logic, and reasoning, as well as where he is in his English language development, from these lengthier speaking samples.

## VIGNETTE: JOSÉ LÓPEZ

José López is a third-grade multilingual learner, originally from Central America, who came to the United States with his family when he was five years old. José was named after his maternal grandfather, whose name was also José, a traditional name in Guatemala and important to his family as it is a name of important people in the Bible such as one of Jacob and Rachel's favorite sons and the husband of Mary, the mother of Jesus. José's teacher frequently organizes students into small groups, or allows some students to choose their partners, in order to provide collaborative mathematical problem-solving opportunities that foster second language acquisition and mathematical content language development. If she feels that individual students need more support than their peers, she selects them to work with her at the front of the class at her table so she can provide them with more individualized attention. José is often selected in such a manner as the teacher says, "José, I would like you to work with me today." José has never been allowed to select his own partner. Most days, after the teacher has finished working with José, she chooses a more mathematically advanced peer to work with José, always reminding these students that they'll need to "go slow, at José's pace."

Based on situations such as these, we began to ask: How can teachers position multilingual learners so they are successful in mathematics classrooms? How might seemingly positive statements and instructional choices create obstacles for multilingual learners? What are specific things that teachers say and do that affect how multilingual learners are positioned?

## WHAT THE RESEARCH SAYS ABOUT POSITIONING

Teachers and students use language to position themselves and others. *Positioning* is a concept drawn from social psychology that explains how people negotiate their identities with others (Harré, 2012; Harré & van Langenhove, 1999; Zangori & Pinnow, 2020). The identities people claim, refute, or are assigned by others are powerful factors in obtaining access to classroom discourse. For example, if students are positioned as incompetent, their contributions may be ignored or discredited. If students are positioned as experts or highly knowledgeable, their contributions will be given greater weight. Thus, through various classroom positioning patterns, students can gain or lose the right to act (Kayi-Aydar, 2015). How students are positioned in your classroom can affect their ability to develop content and language competencies in the classroom. Therefore, positioning is critical to each student's success and learning in the mathematics classroom.

There are three main concepts that constitute positioning: positions, storylines, and acts. *Positions* are social in that they can be viewed as the rights and responsibilities that people are required to carry out, or allowed to carry out, in specific social interactions such as those that occur in the classroom. They are a symbolic notion in that a person may claim a position, or be positioned, along a continuum of categories in any given interaction (e.g., decisive–tentative, extroverted–introverted, competent–incompetent) (Harré & van Langenhove, 1999). *Storylines* can be thought of as narratives about ourselves and others that develop over time in specific communities such as mathematics classrooms (Harré, 2012). *Acts* refer to the social meaning(s) of people's actions (e.g., speech, gestures, gaze) (Harré, 2012; Herbel-Eisenmann, Wagner, Johnson, Suh, & Figueras, 2015). Acts are the means by which we position ourselves and others in interactions. These acts, linked together over time, become storylines about who we are, who we believe others to be, or who they believe we are (our identities and the identities of others).

To make sense of these three concepts (positions, storylines, and acts), think about a time when someone positioned you negatively. Maybe this person labeled you in a particular way (assigned you a position), using spoken language (act), and this initiated a narrative about you in front of others that may have lasted well into the future (storyline). Another example is a child who is described by parents or teachers as "a challenging child" or "a brilliant child." Each of these spoken phrases constitutes an act that positions and introduces a storyline about the child that may be navigated for a very long time by the child and those in the child's environment.

Within each social interaction there may be multiple storylines at play, shaped by people's cultural, historical, and political backgrounds and experiences. For example, a teacher in a classroom may draw on the storylines of reform–traditional instruction and right–wrong mathematics answers simultaneously in interactions with students. This is not to suggest that positioning is always deliberate. Positioning is often inadvertent. Regardless, it is critical to students' success and learning in the mathematics classroom.

How students are positioned in the classroom can affect their ability to develop communicative, social, and academic competencies (Pinnow & Chval, 2014). Although some educators may primarily focus on helping multilingual learners develop English language fluency, they often overlook the importance of social interactions that influence this learning. For example, if a multilingual learner is always partnered with a peer who does all the "talking" and "doing" of the mathematics, then the multilingual learner will be deprived of opportunities to develop communicative competence. Alternatively, if the multilingual learner is always isolated from peers or continually only works with the teacher, then the student is missing opportunities to develop social competencies. Furthermore, if all of this student's activities in the classroom are focused on learning English because it's assumed the student is not "ready" to learn mathematics, then the multilingual learner will be denied opportunities to develop academic competencies—in this case, mathematical competencies (Gibbons, 2015). Therefore, teachers have a pivotal responsibility to productively position multilingual learners. Although some people may think, "Well, I just won't position students," this is impossible. Positioning occurs in every situation. It is not something that cannot occur or can be avoided. With every interaction we are establishing our own identity with others, and often doing things to shape the identities of others, whether we realize it or not. Our words, gestures, silences, gaze, and so on can all convey what we believe about ourselves and others. Thus, in *every* classroom, teachers position students. It is a matter not of *if* a teacher positions, but of *how* the teacher positions.

> *Teachers have a pivotal responsibility to productively position multilingual learners.*

> *In every classroom, teachers position students. It is a matter not of if a teacher positions, but of how the teacher positions.*

## REFLECTING ON XIAO LI'S EXPERIENCE

When we examine Xiao Li's experience, we can see that there is a cultural mismatch between the classroom norms Xiao has been raised to adhere to and the ones her teacher draws upon to shape the norms in this U.S. classroom. Xiao's teacher has categorized Xiao as "shy" because Xiao does not volunteer to answer questions. The teacher is giving Xiao time to adjust, which is an empathetic response. The problem is that Xiao may never volunteer and may take quite a long time to understand that volunteering to speak is not only acceptable in U.S. classrooms, but encouraged. The time lost here is significant for Xiao as opportunities for language and mathematical development can be considerably impacted. The teacher's response is kind; it's just not effective. Xiao needs to be explicitly taught that raising her hand to gain the conversational floor in the classroom is welcomed. Space needs to be made for Xiao to speak in class including the "wait time" (Gibbons, 2015) necessary for Xiao to mentally compose her ideas in her new language. In regard to positioning, the longer Xiao remains silent, and the longer her teacher does not call on her to contribute to class discussions, the more likely Xiao is to be positioned as an invisible student or a silent spectator of classroom events. This position will shape Xiao's storyline in the classroom including how her peers treat her over time (Pinnow & Chval, 2015; Yoon, 2008). In this case, selecting Xiao to speak and teaching her the classroom participation norms are acts that position Xiao in particular ways over time; these are not benign actions that have no consequences for her.

Xiao's teacher also categorizes Xiao as good at mathematics, a student who needs to be "challenged." This is a well-meaning compliment that positions Xiao positively in front of her peers and initiates a storyline of Xiao as successful at mathematics. However, it may rely more on stereotypes than on reality (Chval & Pinnow, 2010). It will be more helpful for Xiao to be assessed in mathematics and have her competencies recognized when they are observed. This will support the teacher as well because she will know who may be strong partners for Xiao in small groups based on data that are accurate.

## STOP AND THINK

Stop and think about Xiao's experience in this situation.

- How would you feel if you were Xiao?
- How can positive positioning actually work against students?

## REFLECTING ON NABIL ABADI'S EXPERIENCE

When we examine Nabil Abadi's experience, we can see that his teacher spends time observing Nabil's interactions first. She does not make assumptions about what might be causing Nabil's frustration, and she does not allow a negative storyline such as "troublemaker" to be instituted. Instead, she observes and notices how particular students dominate small-group talk and activities. Recognizing that this is harmful for every student, she uses the power of her position as the teacher to make space for Nabil to speak. By doing so, she positions Nabil as someone with important contributions to make and one who has ideas that others should listen to and consider. This also provides opportunities for her to learn about Nabil's mathematical sensemaking and give Nabil opportunities to use English to share his mathematical ideas.

A classroom is a busy place, and it is common to overlook such interactions, especially in situations where students may attempt to hide their behavior from the teacher. This is one reason that observing students as they work together and noticing which students self-select to speak more frequently, speak over others, or dominate activities becomes crucial in setting classroom norms that include everyone. It would be easy in Nabil's case to consider his frustration a hallmark of his status and experience as a refugee. However, this thinking can contribute to a storyline and identity of Nabil that exacerbates the issue rather than effectively addressing it. It also will make mathematics teaching and learning more difficult because addressing behavioral issues can become the focus, rather than making room for Nabil and addressing student behavior around equitable and respectful treatment of one another.

## STOP AND THINK

Stop and think about Nabil's experience in this situation.

- How would you feel if you were Nabil?
- How would you feel if your teacher made space for you to talk when others wouldn't listen to you?
- How would you manage the emotions around feeling shut out of participation?

# REFLECTING ON JOSÉ LÓPEZ'S EXPERIENCE

When we examine José López's experience, we can see that José's teacher is trying to support him by working one-on-one with him so that she can facilitate his mathematical problem solving. Over time, though, separating José from his peers can act to position him in particular ways. If individual children are consistently separated from their peers, they will also be deprived of important opportunities to become contributing members of the classroom community.

It is also important to consider the way José is being positioned when the teacher encourages his peers to "go slow" for him. This kind of categorization of José positions him as possibly intellectually slow and incompetent, but we have no data to indicate that this is true about him. It becomes quite important, then, to consider how we choose partners for multilingual learners and how words can shape how others view these learners. A more effective way of positioning José would be to say, "José, I would like you to share your work with the class," compared to "José, I would like you to work with me today."

In each of these examples, José is positioned very differently by the teacher. In the first example, José is positioned by the teacher as someone who can speak for himself and present his own ideas to others. In the second example, José is positioned by the teacher as someone who needs special help rather than as someone who can solve problems through his own efforts. This is not to suggest that teachers helping multilingual learners is wrong or should be avoided. Rather, we want to consider how our organization of learners in the classroom affects how those learners are positioned overall. For instance, allowing some learners to choose their own partners while multilingual learners are categorized through the teacher's talk as those needing "help," or treated as a special project that only involves the teacher, can foster a classroom ecology that positions some students as more competent and capable than others. Students very quickly learn who is positioned in particular ways and will act on that knowledge, which is how particular storylines come into being in classrooms (Pinnow & Chval, 2015; Yoon, 2008). Sometimes teachers in our experience have mentioned that they are concerned about calling on multilingual learners to share their mathematical ideas because it may take them longer to compose and speak their answers in English. However, we have found that making space for multilingual learners to share their ideas positions them as contributors to the classroom ecology, which shapes how their peers treat them. It also shapes their mathematical identity, which is vital to participating in ways that foster both mathematical learning and second language learning.

## STOP AND THINK

Stop and think about José's experience.

- How would you feel if you were José?

- How would you feel if you realized the other kids saw you as a "special case"?

- How would you manage the emotions around feeling separate from others in class?

## POSITIONING STUDENTS AS EXPERTS

Imagine you are a third-grade teacher and you ask a child a question, but the child responds by shaking his head. You may not think too much about it and call on another student, but let's consider this incident from the perspective of a multilingual child. In Transcript 2.1 involving Alonso, a multilingual learner, and Ms. Bristow, his third-grade teacher, you will see one example of how Ms. Bristow responds to Alonso's fear.

On October 14, Ms. Bristow approached Alonso while he was working to ask him to share his ideas on solving a problem—"Today our class has 17 box tops in the basket. Ms. Morales's class has 9 box tops. How many more box tops does our class have?"—during the whole-class discussion at the end of the lesson. Transcript 2.1 occurs immediately after the class has transitioned to the carpet for the discussion.

**Transcript 2.1**

| | |
|---|---|
| Ms. Bristow: | [*Begins by asking students to raise their hands if they used a similar strategy to the one just discussed*] Okay, other people that thought that that would be a strategy for them? Because I saw some friends doing that. Okay. Um . . . Alonso, I am curious about you . . . |
| Alonso: | [*Shakes head vigorously to say, "No"*] |
| Ms. Bristow: | Because you started off thinking—having a similar strategy as Yasmin. Can you tell me about what you thought? |
| Alonso: | [*Whimpers and shakes head no*] |
| Ms. Bristow: | Alonso. [*Smiles at him*] The first thing that he—I'll tell you what Alonso told me. And then, Alonso, you tell me how you figured it out differently. Does that work? Okay. Alonso was thinking along the same lines as Yasmin. He was thinking, "You know, I have 17 box tops, and then I'm going to take 9 more, and I am going to add those together to get 26." But then when he was working with the counters, his ideas changed a bit, and, so, what did you think about the next time that you did it? |
| Alonso: | [*Looks down and moans*] |
| Ms. Bristow: | You're okay. What'd you think the next time? Because a lot of us had a lot of different ideas about this problem and a lot of conversation about it. |
| Alonso: | Mm. That when I add the 17 to the 9 it would equal more than, um, uh—I forgot . . . |
| Ms. Bristow: | More than . . . the box tops that we had? |
| Alonso: | [*Nods head yes*] |
| Ms. Bristow: | Okay. Would that make sense if we started with 17 box tops—would it make sense that we would be getting more box tops? So Alonso noticed that, and then he changed his thinking. |

## STOP AND THINK

Stop and think about what Ms. Bristow says.

- What did you notice about Alonso's reactions?

- How do you think Alonso might have been feeling in the moment?

In the transcript, we can see Ms. Bristow taking the initiative to call on Alonso in front of the class to share his ideas. However, Ms. Bristow did not call on Alonso out of the blue. She asked him in advance if he would share his thinking with his peers. We can see in the transcript that once he was asked in front of the class, Alonso experienced much fear as evidenced by his whimpering, moaning, vigorous head shaking, and reluctance to speak.

▶ As Ms. Bristow pressed Alonso about this thinking, she provided him support in sharing his strategy.

▶ She did not give up and move to another student because Alonso was afraid.

▶ She knew he had a strong answer and needed support to share his ideas with his peers.

Transcript 2.1 highlights the need for teachers to support multilingual learners' participation in whole-class discussions and to not allow multilingual learners to be spectators (Brenner, 1998; Lipka, Sharp, Brenner, Yanez, & Sharp, 2005).

## One Year Later

One year later, this classroom episode was still in Ms. Bristow's mind. During an interview, she shared:

> **“** *If he could have had something to cover himself up, he would have. Last year I didn't want to ask these children [the multilingual learners]. I would have said, "That's fine, you stay hiding." I just refuse to do that now. So, I am going to be as encouraging as I can, [and] I cannot let him do that. So, if he's got something right, we are going to maximize his experiences, and he's going to have to come out of hiding.* **”**

Ms. Bristow notes that she wants to "maximize his experiences." You are also maximizing the experiences of other students in the class when you intentionally plan and include multilingual voices and ideas in the classroom community. Since all students have different problem-solving strategies and ideas, drawing on those ideas as a community benefits every learner. In this way, you position multilingual learners as active participants in the classroom, which influences how their peers position them as well (Turner, Dominguez, Maldonado, & Empson, 2013; Yoon, 2008).

Two months after Transcript 2.1, Alonso was asked to present in front of his peers. Ms. Bristow scanned Alonso's work and projected it on the board (see Figure 2.1).

**Figure 2.1**  Alonso's Problem-Solving Strategies

1.  There are 3 baskets of puppies. There are 19 puppies in each basket.
How many puppies were there in all? 5 7

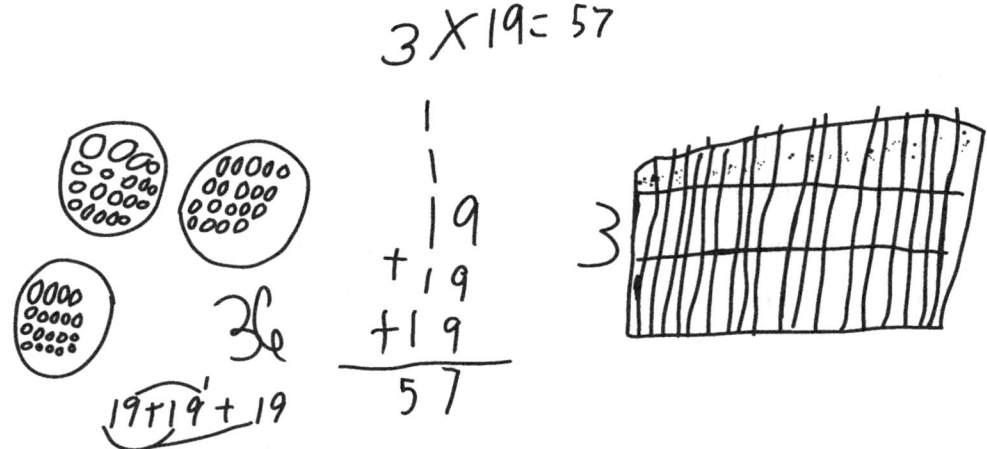

*Source:* Chval, K. B., Pinnow, R. J., & Thomas, A. (2015). *Mathematics Education Research Journal.*
Used with permission.

**Transcript 2.2**

| | |
|---|---|
| Ms. Bristow: | You all had a long conversation about this one. Didn't you? |
| Class: | Yes. |
| Ms. Bristow: | I scanned in Alonso's work because I thought that he did a nice job of explaining this a few different ways. [*Turns to Alonso*] Do you mind showing us one way to explain it for us? [*Addresses the class*] And then he might share other ideas because he did an awesome job of explaining it to Carly. |
| Alonso: | [*Gets up in front of the class*] |
| Ms. Bristow: | If you look, he solved it many different ways. And every single time he solved it, he got the same answer, so that made him even more confident that that was the right answer. |
| Alonso: | On these there was 19. There's 19 here. 19 here. 19. And I counted all of these. |
| Ms. Bristow: | Yeah. |
| Alonso: | And it equaled 57. |
| Ms. Bristow: | Did anyone use a strategy like Alonso? The first strategy Alonso used where you drew a picture of the baskets and put the 19 puppies in the baskets? How many friends did something like that? Okay. All right. Then what's another way you did it? What's that way where you added it up in the middle? Where you were explaining to Carly and I? |
| Alonso: | Oh. This one Carly got so confused that I had to help her on this one because she thought it was 102. |
| Ms. Bristow: | Sure. |
| Alonso: | I took these two and I got . . . 18. And I put 1 up here because there's already a 10. So I put 10 up here and I take this 9 up here and I put it with the 8. And it equals another 10, so I put a 1 up there. |
| Ms. Bristow: | So, you have a group of ten and then seven ones that were left over—seven loose ones—and then you had two rolls or two tens, right? |
| Alonso: | [*Nods*] |

| | |
|---|---|
| Ms. Bristow: | And then you add your tens up? |
| Alonso: | [*Nods*] |
| Ms. Bristow: | Okay, so then you ended up with what? |
| Alonso: | Fifty-seven. |
| Ms. Bristow: | Fifty-seven. So how many friends did something where they added three groups of 19: Did anyone do that? Some of us did that? I see a lot of groups that added 19. Alonso did an array, too, even. Did anyone do an array? Alonso—he did three different ways to solve the same problem. That's a great way to make sure that you got the correct answer, huh? Or four different ways to solve it. Do you see the one over here? |
| Alonso: | Five! Five! |
| Ms. Bristow: | Five! You did the multiplication sentence. |
| Class: | Ooh! |
| Ms. Bristow: | So, if you look, he has that multiplication: three groups, or 3 times 19 is 57. Did anyone else have that same multiplication sentence on their work? You did? Alonso, I was curious about this. Why didn't you do this and do . . . [*Writes on the board*] How come you chose to add the three groups of 19 like that instead of 3 plus 3 plus 3 plus 3 [*referring to adding 19 groups of 3*]? |
| Alonso: | Take me a long time. |
| Ms. Bristow: | That would take you kind of a long time to add all the threes? Some of us did do that. How many friends did that? That's a fine thing if you wanted to add up all the threes. But sometimes, if you wanted to be more efficient, you might think . . . |

*Source:* Chval, K. B., Pinnow, R. J., & Thomas, A. (2015). *Mathematics Education Research Journal*. Used with permission.

## STOP AND THINK

Stop and think about what Ms. Bristow says in Transcript 2.2, compared to Transcript 2.1.

- What differences do you notice between Transcripts 2.2 and 2.1?

- How do Alonso's feelings and actions in this moment compare to those in Transcript 2.1?

In Transcript 2.2, you may have noticed that Ms. Bristow *scanned Alonso's work* so he could use gestures to reference it as he communicated how he solved the problem. This positioned Alonso in a situation as an expert, validated Alonso's work, and functioned to make Alonso feel valued and competent. Moreover, you may have noticed that Alonso's language proficiencies were not a barrier to his participation. This classroom interaction occurred because of Ms. Bristow. Since October (Transcript 2.1), Ms. Bristow *continued to encourage Alonso* to present in front of his peers. She did not allow the October event to mark the end of his opportunities to present in front of the class. Instead, she *recognized Alonso's anxiety and his capability* to push past his fears to present his work in a public space.

## REFLECTING ON WHY POSITIONING MATTERS

Sometimes teachers inadvertently make statements that position students as incapable. In Try It! 2.1, think about the storylines that may result from the teacher's statements about Roberto.

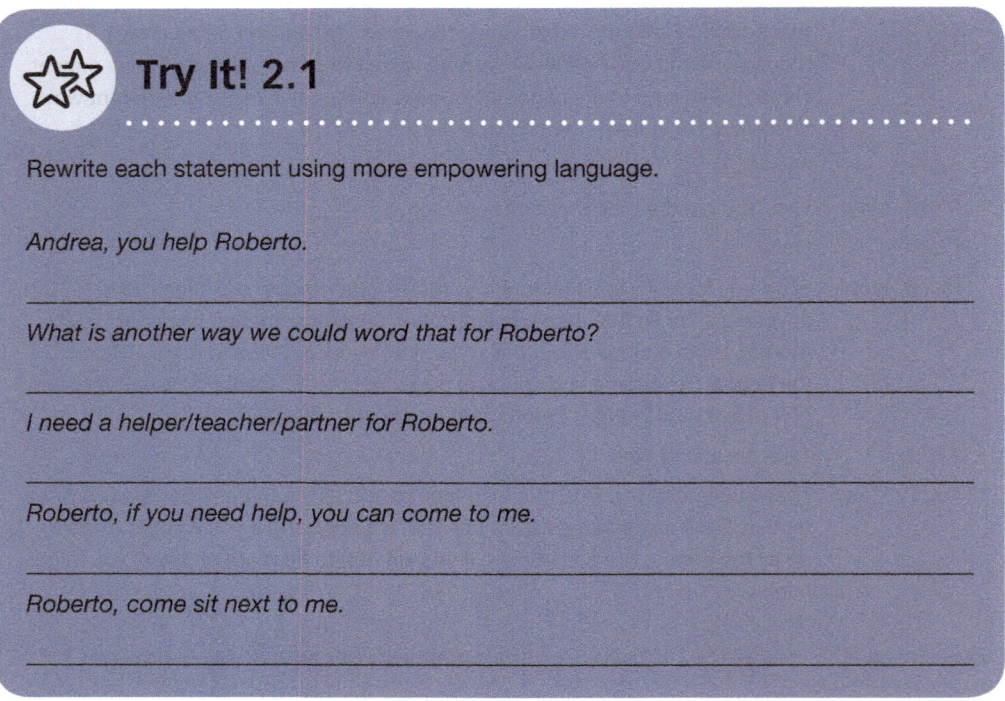

**Try It! 2.1**

Rewrite each statement using more empowering language.

*Andrea, you help Roberto.*

_____

*What is another way we could word that for Roberto?*

_____

*I need a helper/teacher/partner for Roberto.*

_____

*Roberto, if you need help, you can come to me.*

_____

*Roberto, come sit next to me.*

_____

In contrast to how Ms. Bristow positioned Alonso in front of his peers, each of these common teacher statements positions Roberto as incapable. Although unintentional, these statements highlight how teachers think about multilingual learners and influence the way they interact with them. These statements will also affect how Roberto's peers view him and the storylines that develop.

## STOP AND THINK

Stop and think about what you say and do in the classroom.

- What messages are you conveying about multilingual learners?

When multilingual learners are positioned as powerless or troublesome, they become isolated and less interactive in the classroom. The failure to include multilingual learners in meaningful ways in class discussions can serve to position these students negatively over time. Some teachers note that they are afraid to call on multilingual learners in front of others because they are worried this might embarrass learners. However, Yoon (2008, 2012) has noted that, contrary to popular belief, the English language proficiency levels of multilingual learners could not explain multilingual learner participatory behavior across classroom contexts. Rather, when teachers positioned multilingual learners for academic success, multilingual learners engaged in the classroom rather than withdrawing into silence and isolation (Yoon, 2008). More specifically, in Yoon's (2007, 2008) study, multilingual learners who were positioned as powerful by their teacher saw growth in their class participation and were treated more positively by their native-English-speaking peers, regardless of language proficiency. Consequently, it is critical for teachers to understand what positioning is and how teacher positioning can influence multilingual learners' mathematical learning.

## RECOGNIZING HOW TEACHERS POSITION STUDENTS

A critical step to position multilingual learners for mathematical success is recognizing what assumptions you have about multilingual learners and how these assumptions can affect your classroom interactions. As we work with teachers, we are continually inspired by their courage to be honest with us and share their thinking as they work to shift to an asset-based perspective since we are all learning how to do this work together. As we worked with Ms. Jessica Barnes, a third-grade teacher, she was in the early stages of this shift and was working to examine her own assumptions. In an interview, she told us,

> 66 *They [the multilingual learners in her class] are scared to ask questions. They're not confident to come ask. Carlos is the first one that has ever come to me and asked, "What does this word mean, Ms. Barnes?" They are kind of timid, and they stay back. They wait for you to come to them. Sometimes it's hard to remember that some of those words that they don't know—I would have never guessed they didn't know those.* 99

At the start of our work together, Ms. Barnes drew heavily on assumptions that affected the ways she talked about and interacted with multilingual learners and their families. Ms. Barnes didn't notice this at first. Over time, she became aware of how her assumptions about students and her own fear impacted her classroom. Ms. Barnes then began to realize *she* needed to make changes. It's likely the changes she noticed in students' behavior were a direct result of the changes *she was making* in her own thinking. For example, she was now focused on trying to understand what each multilingual learner in her classroom understood about mathematics, instead of creating an environment where multilingual learners didn't feel valued or want to share their thinking.

Positioning starts in how we as teachers view others. If teachers have a deficit perspective, a limited view of multilingual learners' potential, this thinking will be reflected in their words and actions. However, it's important to recognize that what we *don't* say or *don't* do can be just as powerful as overt messages. For example, in classrooms

*What we don't say or don't do can be just as powerful as overt messages.*

*When teachers have an asset-based perspective, they recognize that multilingual learners are competent, bring strengths, and make contributions in classrooms.*

where multilingual learners are allowed to be spectators and positioned as students who lack resources and skills to fully engage, their mathematical development is stunted (Brenner, 1998). Importantly, this student silence is often the result of unfair or inequitable positioning by the teacher, or peers, in mathematics classrooms—not the multilingual learner. When teachers have an asset-based perspective, they recognize that multilingual learners are competent, bring strengths, and make contributions in classrooms.

## POSITIONING MULTILINGUAL LEARNERS AS LEADERS

Up to this point, we've discussed positioning more broadly. For instance, you considered what positioning is and how it can impact multilingual learners in mathematics. Now, we're going to shift our focus specifically to how you position multilingual learners as leaders in your classroom since this is critical for mathematical success. To illustrate how to do this, peek briefly inside two classrooms: that of Ms. Martínez, a fifth-grade teacher who is fluent in English and Spanish, and Ms. Bristow's third-grade class. Their classroom compositions were quite different. Ms. Martínez taught in an urban school where 100% of her students were multilingual learners. Ms. Bristow taught in a primarily white district in a small city. In the year in which the transcript transpired, there were three multilingual learners in Ms. Bristow's grade level, all assigned to her classroom. Both teachers established conditions for success for multilingual learners as individuals, in pairs, and in the whole class by

- eliminating peer domination,
- using strategic partnering,
- setting high expectations, and
- facilitating mathematics and language learning through carefully crafted mathematics lessons.

### Explicitly Positioning Multilingual Learners as Leaders

In Ms. Martínez's fifth-grade classroom, where every student was a multilingual learner, she positioned students as contributors, family members, teachers, role models, and experts through her words and actions. At times, she made these positions explicit through her expectations, such as when she stated,

> *So Dalia was asking Alejandro, and Alejandro didn't know what to do. Alejandro wasn't participating because he never asked for help. So, somebody over here. Anybody. You move around. I'm only one person. Move around quietly and ask each other. You can teach each other. Walk around. Help each other. I can't help all of you at the same time. (Chval, Pinnow, Smith, & Rojas Perez, 2018, p. 123)*

Ms. Martínez clearly communicates that in her classroom multilingual learners are not spectators and can teach their peers. This took place at the beginning of the school year. It was so interesting to watch how her students responded. They were quite surprised by this new norm that they had never experienced. Historically, in earlier grade levels, the students were expected to stay seated, but now Ms. Martínez gave them more powerful positions as teachers in her classroom.

In Transcript 2.3 (from Smith, 2018), you'll peek inside Ms. Bristow's classroom. In this transcript, the class is sitting on the carpet while Lorena, a multilingual learner, shares her way of approaching the problem. As you read, consider the ways Ms. Bristow positioned Lorena for mathematical success.

**Transcript 2.3**

| TRANSCRIPT | POSITIONING ACTS |
|---|---|
| Ms. Bristow: You know, I saw some kids who did a much better job than I did drawing efficient pictures. So, I wanted to talk to you—I wanted a few of those kids to come up. Lorena, you're my first friend to come and share. We're going to talk about number two. Ms. Bristow gave six pieces of candy to Jake, Avery, Carl, and Erica. How much candy did she give out all together? Tell us about your picture. | Positioned Lorena as an efficient drawer in front of her peers<br><br>Positioned Lorena as a student who can explain her thinking to peers |
| *Lorena's work is shown on the board. She has the following drawn on her paper:* <br><br> $6 + 6 + 6 + 6 = 24$ | Scanned Lorena's work so she could use gestures to enhance her explanation of her strategy |
| Lorena: Well, first I made four groups that have six . . . I did 6 plus 6 plus 6 plus 6 equals 12, I mean 24. And then I added. I had to draw a picture of six and then I added them and . . . | Did not interrupt Lorena as she self-corrected when describing her strategy |
| Ms. Bristow: So, um, your picture—did it take very long for you to draw that picture? | Asked Lorena to reflect on the efficiency of her picture in front of the class |
| Lorena: [*Shakes head no*] | |
| Ms. Bristow: No. And you were able to quickly count that there were 24 of them? Wonderful. That's very efficient. Do you guys have comments or compliments for Lorena? | Repositioned Lorena as an efficient drawer in front of her peers; expected peers to attend to Lorena's mathematical thinking |

*Source:* Smith (2018).

## Implicitly Positioning Multilingual Learners as Leaders

In Transcript 2.4, Lorena is working in a small group at the "kidney bean table." Ms. Bristow stops the class and asks them to approach Lorena so that she can explain her thinking. In this case, Ms. Bristow did not have students come to the carpet as she typically did. Instead, Ms. Bristow stopped everyone and made the class come to Lorena. By doing this, Ms. Bristow positioned Lorena as an expert her peers needed to listen to. As you read, you'll notice a number of other students being called on. Keep in mind that only Lorena and Linda were in Lorena's group and were asked to share their thinking. The other students are being called on in the conversation to ensure they are being attentive to Lorena's explanation.

**Transcript 2.4**

| | |
|---|---|
| Ms. Bristow: | Okay, so Lorena, can I borrow these? What they did was when I came over here each person had a little stack of these [*holds up two base ten boxes*] in front of them. |
| Lorena: | Like, like this [*places boxes on table to illustrate to class*]. So . . . |
| Ms. Bristow: | So like if— |
| Lorena: | Mine was 212. I have . . . |
| Ms. Bristow: | [*Whispering*] Sorry to both you guys. Can you guys listen? [*Normal voice*] Her number was 212. |
| Lorena: | [*Moves base ten blocks on table*] |
| Ms. Bristow: | She had two boxes, one rod, and— |
| Lorena: | Twelve ones. |
| Ms. Bristow: | Okay, so do you need twelve ones and a ten? |
| Lorena: | No, no. |
| Ms. Bristow: | So, you have two hundreds, one ten, and how many ones do you need for 212? Have you written it down? |
| Lorena: | I have 12. |
| Ms. Bristow: | Where's yours? Let's write it. Oh here, see, you have two boxes— |
| Lorena: | And one . . . |
| Ms. Bristow: | One roll and . . . |
| Lorena: | Twelve—and two . . . |
| Ms. Bristow: | And how many loose ones would you need? Okay, so Lorena has her stack of shirts. Okay, what do you think this represents, Keri? [*Holds up one block*] |
| Keri: | A hundred. |
| Ms. Bristow: | Or? . . . How many boxes? |
| Keri: | One. |
| Ms. Bristow: | One box. So, she has her two boxes, her one roll, and her two . . . What are these [*holds up two single cubes*], Stephen? |
| Stephen: | Ones. |
| Ms. Bristow: | What are these guys? [*Holds up two single cubes*] |
| Stephen: | Ones. |
| Ms. Bristow: | Or we call them loose shirts, right? Bruce, come on over. Jack, come on over by me. All right, so then each person in her group also had—had a set of shirts in front of them, right? So, then what are you doing now? |

| Lorena: | Now we were adding them up. So, we made one group of hundreds and then one group of tens and one group of ones and then we added them all together. Linda said she had . . . |
|---|---|
| Linda: | I had 164. So, I did this and then six of these. |
| Ms. Bristow: | Keri, come stand by me, please. Cory, come stand by me. |
| Linda: | Six of these and four ones, so we each have those in front of us, and then what we would do after we had them in front of us, we would have a pile of tens . . . |
| Ms. Bristow: | [*Low tone*] Come here. |
| Linda: | And then we would have our pile of ones. |
| Ms. Bristow: | [*Talks to a student*] You need to participate in this. |
| Linda: | And our pile of hundreds, and so we would add them up together and we are going to do that, and then we are going to try to find out our final number by adding up all of them. |
| Ms. Bristow: | So, you know, I really like this group's strategy, 'cause I'm seeing that some of us are feeling maybe a little frustrated not knowing exactly which tool would be useful—do you guys think these place value blocks are working for you pretty well today? |
| Students: | Yeah. |
| Ms. Bristow: | Is that a strategy you guys might try—Lorena's strategy—when you guys get back to your tables? |
| Students: | Yeah. |
| Ms. Bristow: | It might be a useful strategy for many of you who maybe are kind of at a point where you're not quite sure what to do—this might be a really good strategy for you guys. Lorena, thank you so much for sharing your thinking. [*claps*] Awesome strategy. Okay, go on back and get started on that. |

In Transcript 2.4, you can see Ms. Bristow addressing the following unproductive practices (Figure 2.2).

**Figure 2.2** Recap of Transcript 2.4

| UNPRODUCTIVE PRACTICE | MS. BRISTOW'S ACTS |
|---|---|
| Silence, which can facilitate neglect of multilingual learners | Calling the class to Lorena's group and asking Lorena to share her strategy |
| Others talking for, or over, multilingual learners | Quickly refocusing students' attention back to Lorena when students became disengaged or disrespectful |
| Not giving multilingual learners the space and time needed to process their ideas and speech | Allowing Lorena time to process her thoughts |
| Always "rescuing" multilingual learners | Allowing Lorena to self-correct when presenting her work in front of the class |

# STRATEGIES FOR POSITIONING MULTILINGUAL LEARNERS AS LEADERS

Here are some strategies you can use to position multilingual learners as leaders in your classroom (see also Try It! 2.2):

▶ Ask multilingual learners to share their mathematical thinking with the class during discussions. Ensure all students are respectful listeners as the student presents.

▶ Have multilingual learners present their ideas at the board. Scan their work so it can serve as a visual referent.

▶ Encourage multilingual learners to "teach" peers.

▶ Assign ownership of mathematical ideas or strategies to multilingual learners, such as "Lorena's strategy."

▶ Point out what is valuable about a multilingual learner's idea, strategy, approach, or representation—for instance, "Marco was being very efficient," or "If you want to be more efficient, you might use Esme's strategy."

▶ Give multilingual learners time and space to process and communicate their thoughts.

Sometimes the multilingual learners in your class may not yet possess the competencies to communicate their mathematical thinking in English. In such cases, you may

▶ ask multilingual learners to share their mathematical thinking in their first language (as opposed to English), which demonstrates to the class that these students have mathematical ideas worth sharing;

▶ ask multilingual learners to read in their first language if they have this competency, which shows a different aspect of these students to peers—specifically that they can read and know things;

▶ ask students who have learned words and phrases in different languages to share their knowledge and language with the class, which positions multilingualism as valued; and

▶ provide multilingual learners time to share their ideas with you as the teacher before sharing with the class as a whole as this provides space and time to practice speaking their ideas aloud to others.

# THINKING ABOUT POSITIONING IN YOUR PRACTICE

In this chapter, you looked into the classrooms of Ms. Martínez's fifth graders and Ms. Bristow's third graders to examine the different ways they positioned multilingual learners in their classrooms for success. By examining their practice, you saw Ms. Martínez and Ms. Bristow position multilingual learners as leaders in the classroom, both explicitly and implicitly. Overall, these positionings changed the participatory behavior of multilingual learners and the ways they were treated by their native-English-speaking peers. Both teachers changed the storylines of multilingual learners in their schools when they positioned them as leaders.

### Try It! 2.2

Select one of the strategies listed in this chapter that you could use to position multilingual learners positively. Try the strategy in your classroom each day for the next week. After each class period, reflect using the following questions as a guide:

- How did you implement the strategy? What did you say and do? When did you try it in the lesson?

- What did you notice about how the student(s) responded?

- What is one thing you will do when you try the strategy again?

Contrary to popular belief, student silence is often the result of unfair or inequitable positioning in content classrooms (Pappamihiel, 2002; Pinnow & Chval, 2015). Such positionings can subsequently reduce multilingual learners' opportunities to engage in meaningful learning experiences. Pappamihiel found:

▶ Mainstream students dominated almost every interaction with multilingual learners. For example, mainstream students grabbed the writing implements and paper, computer mouse, and learning tools and placed them in front of themselves, keeping the multilingual learners from using them (i.e., moving own pieces). Mainstream students dominated talk—both content and social talk.

▶ Multilingual learners are frequently left to navigate openings in positioning on their own.

Consequently, when multilingual learners are positioned inequitably in peer-to-peer and whole-class interactions, it can make it even more challenging for multilingual learners to gain access to academic debate and discussion. According to Chval and Pinnow (2010, p. 10),

> Teachers must (1) promote active ELL [English language learner] participation in mathematical discussions, and (2) recognize the resources that ELLs use to express mathematical ideas in order to facilitate participation and learning of ELLs, especially Latinx ELLs in mathematics classrooms (Brenner, 1998; Brown et al., 1993; Khisty & Chval, 2002; Moschkovich, 2002).

## Reflect

- Imagine you were to record and study your teaching. What would you find about your positioning of multilingual learners?

- What strategies will you use more often to position multilingual learners as leaders in your classroom?

- What storylines are present in your mathematics classroom? Are there storylines you want to alter? If so, how will you alter them or promote new ones?

# CHAPTER 3
# FACILITATE MULTILINGUAL LEARNERS' PARTICIPATION IN MATHEMATICS

## Key Concepts

In this chapter, you will

- ✓ identify different factors that influence the participation of multilingual learners.

- ✓ identify strategies that facilitate the participation of multilingual learners.

Throughout our teaching careers, we have experienced situations where students were reluctant to participate in classrooms, and we respected their silence when appropriate. Each of us may also have experienced this type of reluctance in our own professional settings. In other words, we can all be reserved in different contexts. In contrast, we've also experienced situations where everyone was eager to speak up. These contrasting experiences facilitated our desire to consider questions—for example, how can teachers encourage participation from *every* student? How can students encourage the participation of their peers? While working with teachers in classrooms, we have observed multilingual learners who did not participate in their mathematics lessons. As discussed in Chapter 2, we do not want students to be silent spectators, because we know that a lack of participation hinders their mathematics and language development. Therefore, we must identify specific factors that influence participation, and implement strategies that facilitate participation for every child.

## REFLECTING ON YOUR EXPERIENCES WITH PARTICIPATION

In your professional roles, you have the opportunity to participate in various settings outside of your classroom, such as faculty or committee meetings, conversations in the hallway, professional development sessions, and conferences with parents. Think about your participation in those different settings. What are factors that influence your participation? In other words, when do you feel comfortable sharing your ideas and thinking? What would hinder your participation in a meeting, professional development session, or conversation with a colleague?

You may feel that your participation is hindered depending on the context. For example, you may feel comfortable sharing your thinking while you have coffee with your grade-level team; however, when all of the faculty in the school are together to discuss a challenging issue, you prefer not to stand up in front of the group. Your participation may be hindered if you do not know the participants in the room or you have a previous history with someone in the room that has been challenging. In the case of a professional development session related to mathematics, your participation may be hindered if the instructions are unclear, the content is unfamiliar, or you feel there are experts in the room who are more knowledgeable or have more years of teaching experience. Now imagine attempting to participate in each of these situations in a nonnative language.

In many U.S. schools, English is the primary language of instruction. Multilingual learners navigate classroom participation structures in a nonnative language, often translating ideas and questions from English into their first language (or dominant first language in the case of learners who speak several languages in addition to English such as students from Tanzania or India), building an internal response, and then translating their answer back into English to share with others. This internal work can take time as multilingual learners process their mathematical thinking and the language involved in expressing that thinking. Multilingual learners in instances like these are often not given sufficient "wait time" (Gibbons, 2015) in which to complete this work, which can position them as learners without important ideas to add to the discussion. It is important to consider what messages this sends to others about the value of multilingual learners' contributions to class work if they are not allowed to create and convey their ideas.

Some of the reasons that hinder your participation may be different from those that hinder the participation of students, especially multilingual learners. Think about a student you have observed whose participation was hindered. Focus on the image of that student. Based on your observations or interactions with this student, what are factors that hindered this student's participation?

## STOP AND THINK

Stop and think about what hinders your participation as an adult and what potentially hinders the participation of students, especially multilingual learners.

- How do your answers compare? In other words, what is the same or different about the factors you identified?

# WHAT THE RESEARCH SAYS ABOUT STUDENT PARTICIPATION

Current curriculum and instructional recommendations emphasize cognitively demanding mathematical learning tasks for all students, language-rich environments, and multiple modes of communication (National Council of Teachers of Mathematics, 2000). Debates suggest that these recommendations may further disadvantage students from low-income and underrepresented groups and widen existing educational and social disparities (Jhagroo, 2015; Lubienski, 2000, 2002; Secada & De La Cruz, 1996), especially if students are silent nonparticipants in the classroom. However, when instruction facilitates full engagement and participation, multilingual learners are successful (Pinnow & Chval, 2015; Razfar, Khisty, & Chval, 2011).

Student participation is a crucial component for the mathematical learning of every student. Since students learn not only from the teacher, but also from one another, it is important that everyone participates in the classroom. Participation is also important because it can provide language development for every learner including multilingual learners who are learning mathematics and the English language (Gibbons, 2015). To facilitate participation of multilingual learners in mathematics classrooms, teachers must provide opportunities, promote active student participation in mathematical discussions, and recognize the resources that multilingual learners use to express mathematical ideas (Brenner, 1998; Brown et al., 1993; Moschkovich, 2002, 2013, 2015; Turner, Dominguez, Maldonado, & Empson, 2013; Zahner, 2012).

Through social interactions over a sustained period of time, students come to share common ways of thinking and communicating. During the social activity in a mathematics classroom, students encounter language focused on mathematical tasks, which is spoken, written, drawn, and gestured. The context and nature of that discourse, including the nature of the mathematical tasks, influences what students learn (Henningsen & Stein, 1997). Examining learning involves considering the practices, tools, and interactions within a community of practice (Hansen-Thomas, 2009; Lave & Wenger, 1991; Takeuchi, 2015). Lave and Wenger (1991) wrote:

> 66 By this we mean to draw attention to the point that learners inevitably participate in communities of practitioners and the mastery of knowledge and skill requires newcomers to move toward full participation in the sociocultural practices of a community. "Legitimate peripheral participation" provides a way to speak about the relations between newcomers and old-timers, and about activities, identities, artifacts, and communities of knowledge and practice. (p. 29) 99

Therefore, it is important to consider factors that may hinder full participation, as well as teaching strategies that will facilitate participation.

## FACTORS THAT INFLUENCE THE PARTICIPATION OF MULTILINGUAL LEARNERS

As you reflected on the student you identified earlier, you may have considered factors that hinder students, such as

- fear of embarrassment or potential ridicule in front of peers;
- fear of the teacher's reaction to a contribution;
- lack of knowledge or competency related to content;
- expectations (e.g., in some cultures, students are not expected to speak unless called on by the teacher);
- classroom norms (e.g., some students dominate or jump in, or disregard or disrespect others' contributions); and
- experiences (e.g., a student had an argument with a parent prior to class).

No matter what the reasons are, these factors are problematic because they hinder student participation and learning.

### Reflecting on Your Experiences With Fear

Have you ever felt like an outsider at a new job or place, when you wanted to join a group, when you arrived late to a meeting, when you were the only person who didn't look or talk like others, when you didn't understand a joke, when you visited a foreign country or hung out with teenagers, or when you didn't know the lingo of a group? You may have also felt like an outsider in professional spaces, such as at faculty meetings, during a professional development session, or on the first day of class. Regardless of the situation, when you feel out of place, you may not feel valued or respected. In addition, you may experience reduced self-confidence and/or doubt your own abilities. Such feelings can lead people to be silent, disengage or not participate, avoid the situation, pretend they're someone they are not, go into survival mode, or hide. Thus, it's important to recognize:

▶ People experience the feeling of being an outsider in different situations.

▶ Everyone experiences fear.

▶ There are different ways of acting and behaving that occur in response to feelings of fear and anxiety.

▶ Your perception of how others perceive you influences the ways you act and behave.

Students have the same experiences and emotions, which influences their learning. Therefore, it's important to consider fears students may experience in classrooms. To do this, we will begin with a general discussion of student fears before we consider multilingual learners, specifically.

## Overcoming Fear

Watch Video 3.1 (Jukin Media, 2012) showing a young girl named Zia who is attempting her first large ski jump. As you watch the video, look for evidence of Zia's anxiety as she considers making the jump.

**VIDEO 3.1:**

Jukin Media (2012, March 12). *Girls first ski jump.* YouTube. https://youtu.be/ebtGRvP3ILg

To read a QR code, you must have a smartphone or tablet with a camera. We recommend that you download a QR code reader app that is made specifically for your phone or tablet brand.

## STOP AND THINK

Stop and think about Zia's fear.

● Is Zia's fear real?

● What helps Zia overcome her fear?

● How would you characterize the fear some students feel in many mathematics classrooms?

In the video, Zia, a fourth-grade girl, wears a head camera, which captures her self-talk and an interaction with her coach, a more experienced skier. Watch the video a second time, focusing on Zia's self-talk and her coach's response.

● What do you notice about Zia's self-talk?

● What do you notice about what and how Zia's coach responds to her?

It is evident in the video that Zia experiences feelings of fear and anxiety; however, she overcomes these through positive self-talk, reminding herself that she "can do it," and builds on her previous knowledge and experience of jumps. In addition, she consults someone more experienced (e.g., coach, instructor), who encourages her, provides information, remains emotionally calm, and does not give her an out or attempt to "rescue" her. Altogether, these factors facilitate Zia's position as someone who can be successful with the jump.

Just like Zia, you have experienced fear. Sometimes these fears can occur in a classroom. For instance, imagine if a student thought:

> **"** *I'm afraid I won't fit in with my peers.*
>
> *I'm afraid others don't know all that I can do.*
>
> *I'm afraid my English isn't strong enough to share.*
>
> *I'm afraid I'll embarrass myself.* **"**

As a teacher, you may forget that students can be fearful in your classroom. However, you need to consider what fears your students may have and work toward creating an environment that diminishes those fears that can restrict student learning (Allen & Chval, 2009). One way to do this is to have a discussion with students where they share their fears (see Try It! 3.1). In this discussion, it's important that you validate these fears and confirm that everyone may experience them. Then, as a class, create a list of how to respond to each situation when fear may arise and how the fear can be overcome. In this conversation, it's important to help students identify tools or strategies they can use to overcome fear.

*You need to consider what fears your students may have and work toward creating an environment that diminishes those fears that can restrict student learning.*

 **Try It! 3.1**

Ask your students what fears they face in mathematics classrooms.

## Encountering Unknown Contexts

To identify additional factors beyond fear, consider two mathematics problems. What would hinder your participation in solving and discussing the following mathematics problems?

1. Max and Mario are playing jai alai with Priti and Vandana. Max and Mario have 3 points, and Priti and Vandana have 4 points. How many points do Max and Mario need to win? How many points do Priti and Vandana need to win?

2. There are 3 okapi and 4 shoebills. How many legs are there altogether?

*Hint:* Jai alai is a "sport involving a ball that is bounced off a walled space by accelerating it to high speeds with a hand-held wicker device" (Wikipedia, 2020).

As we select or write mathematics problems for students, including multilingual learners, it is important to consider how language may constitute a barrier that cannot be addressed simply by providing a list of definitions. Suppose you provide the students with the definitions of *jai alai, okapi,* and *shoebill.* The definition may not be sufficient for solving the problem. In the problem about jai alai, students need to understand the rules of the game, how to play it, and the context of the game. With the second problem, solvers need to know how many legs are on an okapi and a shoebill.

If you were a multilingual learner as a child, you are in a better position to appreciate and empathize with multilingual learners. As demonstrated using the mathematics problems, mathematics curriculum materials may include problem contexts that are unfamiliar to multilingual learners. As you examine contexts that you use in your classroom, it is productive to learn about the cultures and experiences of your multilingual learners and create contexts familiar to them. This approach honors the cultures of multilingual learners and enhances the lives of their peers who may not be familiar with cultures other than their own. In Chapter 5, you will find a more extensive discussion of contexts, and in Chapter 12, you will learn about enhancing curriculum materials for multilingual learners.

## STOP AND THINK

Stop and think about examples of sports that are found in your mathematics textbooks or curriculum materials. Are there specific contexts that are unknown to multilingual learners?

Some students from other countries may not be familiar with U.S. football or basketball. Sometimes, the use of familiar contexts for U.S. students can constitute a barrier for multilingual learners, when they are not acquainted with these contexts. These challenges are not unique to multilingual learners. For example, as students relocate across the country, they may experience new foods, animals, or landmarks that are unfamiliar. In addition, curriculum materials may include mathematical conventions or mathematical algorithms that differ from what students have learned in their countries of origin.

In the second problem, a photo or picture of an okapi and a shoebill along with the use of gestures would support multilingual learners in building meaning for *okapi, shoebill,* and *leg* in this context. These two simple examples highlight dramatic differences with mathematics problems. In the second problem, a visual representation and gestures may provide access for multilingual learners to solve the problem. However, with the jai alai problem, a picture will not help solvers understand the rules of the game, necessary information for solving the problem.

As these examples show, every context will require you to make decisions about when the context is important and when you will need to invest time to support multilingual learners in building meaning for it. You may decide to change contexts in mathematics problems so that they do not act as barriers to learning mathematics. You will also have to recognize when contexts are not the issue, but rather children have not developed a sufficient mathematics knowledge base to tackle the problem. Therefore, it is important for you to analyze the contexts in your curriculum materials. We will discuss this idea further in Chapters 5 and 12.

*Removing barriers to participation for multilingual learners is critical for their success.*

Removing barriers to participation for multilingual learners is critical for their success. In the remainder of this chapter, you will examine transcripts from classrooms to identify strategies that will eliminate or diminish these factors in your mathematics classrooms.

## EXAMINING TEACHING STRATEGIES THAT INFLUENCE PARTICIPATION

Dr. Chval was working with Ms. Keller, a third-grade teacher, to enhance the participation of multilingual learners in her mathematics classroom. Ms. Keller asked Dr. Chval to lead a lesson about solving word problems. To begin, Dr. Chval wrote two sentences on the board about balloons (i.e., "Dr. Chval has 24 balloons" and "15 balloons popped") and then solicited different questions from third graders that could be asked based on the information.

As you examine Transcript 3.1 from Ms. Keller's classroom, notice

- ▶ how we describe the strategies the teacher uses, and whether you would describe the strategies the same way;

- ▶ how the teacher engages multilingual learners; and

- ▶ how the teacher asks students to create questions to include in the mathematics problem.

As you read, focus on what you can learn rather than on what the teacher should have done differently. The left-hand column contains the transcript, and the right-hand column contains an annotated description of the instructional strategies.

**Transcript 3.1**

| TRANSCRIPT | TEACHING STRATEGIES |
|---|---|
| Dr. Chval: Dr. Chval has 24 balloons [*writes "Dr. Chval has 24 balloons" on the board*]. Does everyone know what a balloon is? | Spoke and wrote information on the board. Asked the class if they were familiar with the word *balloon*. (This could have been improved if Dr. Chval had a physical balloon.) |
| Students: Yes. | |
| Dr. Chval: Yes. Okay. What happens if I take a pin and I stick it [*gestures to pop a balloon*] in the balloon? | |
| Students: It'll pop. | |
| Dr. Chval: It pops and explodes [*gestures an explosion*], right, and you don't have your balloon anymore. | Asked the class about popping balloons and included a gesture of popping a fictitious balloon. Used gestures to mimic a balloon exploding. Repeated the word the children used (i.e., *pops*) and introduced *explodes* to help build meaning for *pops*. Emphasized the balloon goes away to set up the idea of subtraction. |
| Erin: And if you took away that balloon. | |
| Dr. Chval: You're taking away that balloon, [so] you don't have it anymore. So, let's say that 15 balloons popped [*writes "15 balloons popped" on the board*]. | Repeated Erin's contributions to validate them and revoiced them (e.g., "You're taking away that balloon, [so] you don't have it anymore"). Wrote each sentence, which provided a visual referent, and stopped to discuss the first one to ensure understanding before writing the second. Wrote the second sentence using the word *popped* that the students gave her. |
| Michael: So, write "24 minus 15 equals . . ." | |
| Dr. Chval: Okay. Now, you've been doing some problems in here in math, and sometimes you add and sometimes you subtract. How do you decide whether you add the numbers or subtract the numbers? How do you decide that? Yes? | Ignored Michael, who wanted to get to number expression and answer a computation situation. Laid the foundation for a larger mathematical picture in which the emphasis is not on a single problem, but on thinking about the process of decision making for different problem-solving situations. |
| Craig: Sometimes if it says that, like, you have 24, then somebody else gave you more, that means, um, addition. | |
| Dr. Chval: That's addition. | Revoiced *addition* to emphasize. |
| Craig: It's like if, um, you had 24 of them and—and somebody stole 15 of them. | |
| Dr. Chval: Okay, and then you would have less, right? | Focused the class on the idea of having less to set up the concept of subtraction. |
| Craig: Yes. | |
| Dr. Chval: And when we have less than, we subtract. So, that's what you have to think about when you— | |
| Tad: What's he talking about? | |
| Dr. Chval: He said that if you have 15 and you get some more that's going to be a— | Revoiced Craig's contribution. |
| Tad: You lost 15. | |
| Dr. Chval: Let's say you had 15 more than. You're going to have an addition problem, right? Because you're getting more. But if I lose some— | Emphasized the importance of sense making in problem solving. |
| Tad: You lost 15. | |

(continued)

(continued)

| TRANSCRIPT | TEACHING STRATEGIES |
|---|---|
| Dr. Chval: He said—we're trying to decide, *When do we add 15, and when do we subtract 15?* So, when I get a problem, I have to make a decision. You're going to start working with multiplication, and then you're going to have another one where *Do I add, do I subtract, [or] do I multiply?* Right, and so how can I make sense of a problem? How can I figure out what to do with a problem? I have to make sense of it. So, usually you get information, and then there's a question. Who thinks they know what my question is? Okay, so raise your hand if you know it. *Dr. Chval has 24 balloons. 15 balloons popped.* So, what's the question? Dr. Chval wants to know. | Introduced the structure of many word problems (i.e., given information, then a question). Asked students to determine the question rather than giving it to them.

Repeated what she wrote to give students more time to think about a question. Clarified the request for students to generate a question. |
| Melody: The answer. | |
| Dr. Chval: No, I don't want to know what the answer is. I want the question. Ellie? | Clarified that she does not want a numerical answer, but rather is looking for a question. |
| Ellie: 11. | |
| Dr. Chval: You're giving me answers, I don't want a number yet; I want a question. So, what would be my question? *Dr. Chval has 24 balloons. 15 balloons popped.* I want to know what the temperature is outside. No, I don't want to know that. What do I want to know about the balloons? What's the question? Do you have a question? | Repeated what she wrote to give students more time to think about a question. Overall, she repeated the question three times and pointed to words as she read it. Provided an example of a question that is ridiculous. Gave students an opportunity to generate questions, which was hard for them to do since they had not done it before. |
| Leticia: How many popped? | |
| Dr. Chval: How many popped? Excellent question. That's exactly right. How many popped? Who has a different question?

We could write lots of questions. How many balloons popped? [*Writes "How many popped?" on the board*] Raise your hand [*gestures hand raising*]. "How many popped?" could be a question. Who has a different question that we could use? Do you have a different one? [*Points at a student*] No. Do you have a different one? [*Points at another student*] | Revoiced the question and validated the response from Leticia, a multilingual learner, by stating it was an excellent question and highlighting there were no limitations on questions. Reiterated that a multitude of different questions existed that could be asked in this situation. Then documented the question on the board.

Requested additional questions. |
| Tiffany: Yes, um, how, um, many do we have left? | |
| Dr. Chval: How many do we have left? Perfect. That's a great question. How many are left? [*Writes "How many are left?" on the board*] We could ask that question. What's another question we could ask? | Revoiced the question and validated the student's response. Then documented the question on the board and requested additional questions. |
| Marquese: How many does she have now? | |

| TRANSCRIPT | TEACHING STRATEGIES |
|---|---|
| Dr. Chval: How many does she have now? Yes, excellent. So, could you see these questions? Could Ms. Keller give you some questions next week that say how many popped or how many are left? | Revoiced the question and validated the student's response. Then documented the question on the board. |
| Luke: Well, actually, how many popped because it's 15 so that wouldn't really make sense. | |
| Dr. Chval: You don't think that one makes sense? Well, it does because sometimes teachers just want to see if you can read a problem. Right? So, could you read the problem and know, "I'm not supposed to do anything. All I have to do is look for the information." So, what was the question you just gave me, Marquese? | Reiterated the validity of the question, which is significant since a peer, Luke, openly challenged the validity of or attempted to discredit Leticia's contribution. In this situation, Dr. Chval did not allow this to happen and used her authority to validate the multilingual learner's contribution. Moved on, returning to Marquese's response. |
| Marquese: How many does she have now? | |
| Dr. Chval: How many does she have now? [*Writes "How many does she have now?" on the board*] Okay, very good. | Revoiced the question and validated Marquese's response. Then documented the question on the board. |

## STRATEGIES FOR FACILITATING MULTILINGUAL LEARNERS' PARTICIPATION IN MATHEMATICS

In Transcript 3.1, Dr. Chval engages the students in a discussion about a mathematics problem as she asks students to analyze and discuss problem structure; solicits multiple answers to her question; writes notes and the problem on the board as she speaks, repeating the two sentences three times to provide additional time for multilingual learners to participate; uses gestures; and encourages the active participation of a multilingual learner by taking up Leticia's response and validating it in front of everyone. Dr. Chval also does not cede the discussion to Michael when his efforts focus solely on the product of a correct answer rather than a discussion about the process of problem solving.

## STOP AND THINK

Stop and think about what strategies you use to engage multilingual learners and validate their contributions.

Use Try It! 3.2 to analyze Transcript 3.2, an excerpt from a first-grade lesson taught by Ms. Reams. In this lesson, drawn from Chval, Chávez, Pomerenke, and Reams (2009), students are expected to collect attendance data for the day and create a graph using colored paper. To introduce the lesson, Ms. Reams tells a story about an unusual day where most of the students are absent due to extraordinary events that occur on the way to school. For example, some of the students climb into a giant bird's nest and never make it to school. Following the story, Ms. Reams asks the students to represent the attendance data from the fictitious story.

 **Try It! 3.2**

As you read Transcript 3.2, use the right-hand column to make note of teaching strategies you observe. How did Ms. Reams engage multilingual learners?

**Transcript 3.2**

| TRANSCRIPT | TEACHING STRATEGIES |
|---|---|
| Ms. Reams: Who has ever heard of the word *unusual*? What does that mean, *unusual*? . . . Who has ever had an unusual dream? Something that's just kind of weird? I think of the word *weird* when I think *unusual*. Has anyone ever had a weird dream that's not something that happens every day? Hannah, you have? Okay. Thomas, have you had an unusual dream? You have. Catherine, have you? You've had an unusual dream. . . . Last night I had a very unusual dream. It had to do with you. | |
| Students: Tell us it! Tell us it! | |
| Ms. Reams: Are you sure? I'm not sure. | |
| Students: Yes! | |
| Ms. Reams: Are you ready to be listeners? | |
| Students: Yes! | |
| Ms. Reams: Let me see [*that you are ready to be listeners*] because we're going to laugh some, but you have to listen, too. Okay? I'm going to tell the story and be writing stuff as I tell it, and I'm going to call it *The Unusual Day*. [*Writes "The Unusual Day" on the board*] | |
| Student: You mean *night* or *dream*? | |
| Ms. Reams: [*It was*] my dream—I had it. But it was about an unusual day here [*at school*]. *The Unusual Day*. And then if I say your name in this dream, I want you to come put it up, and I'll tell you where to put it up, okay? But you're going to put your magnet up on the side that it belongs, okay? So, these are the people that came to school on the unusual day. They were here. These are the people that didn't come to school on the unusual day. They were not here. | |

**Figure 3.1**   Students examine the photos and images on the board to complete the task.

As the last segment of Transcript 3.2 explains, Ms. Reams uses magnetic name tags of the children to help illustrate the story. She also uses the board to capture the entire storyline. As the children are mentioned in the story, they walk to the board to add their picture, thus enhancing the student participation beyond just listening. For example, when Ms. Reams reads that in her dream Mary, James, and John investigated the bird's nest and didn't make it to school, these three children walk to the board and place their pictures next to the picture of the bird's nest drawn by Ms. Reams. Throughout the lesson, Ms. Reams does not erase the board (see Figure 3.1) so that the children can refer to the board during the story as well as during the assessment that follows.

Ms. Reams used a variety of strategies that are critical to the success of multilingual learners:

- ▶ Connected the details in the story to students' lives
- ▶ Required active participation of students (who placed their name magnets on the board)
- ▶ Emphasized key language (e.g., *unusual*)
- ▶ Incorporated student names from her classroom into the story
- ▶ Drew pictures (of some of the obstacles that inhibit student attendance in the story)

In addition, Ms. Reams enhanced this part of the lesson using other strategies identified in the research about teaching multilingual learners. She replaced the beginning of the story, "Once upon a time . . . ," by saying, "Last night I had a very unusual dream." This adaptation helped Ms. Reams connect the story to students' life experience. She also repeated the meaning of *unusual* so that the children could better understand the new word and therefore fully participate in the conversation.

## THINKING ABOUT PARTICIPATION IN YOUR PRACTICE

In every lesson, teachers consistently make decisions about which teaching strategies to use and when to use them. All of the chapters in this book introduce strategies that will ultimately facilitate the participation of multilingual learners. We see all of these ideas as interdependent and important to consider in designing every mathematics lesson. In addition, the following questions may be helpful to consider while you plan your lessons to enhance participation for multilingual learners. We will continue to delve deeper into these ideas throughout the book.

### Questions to Consider While Planning Lessons

- How can I best utilize the classroom's board/writing space during the lesson?

- How can I connect the mathematics to students' life experience and existing knowledge?

- How can I help children build meaning for the mathematical language involved in the lesson?

- How can I encourage students to use gestures or drawings to communicate their mathematical thinking?

- What are the key mathematical ideas and concepts that need to be emphasized during the lesson?

- What representations would help the children build meaning for these key mathematical ideas and concepts?

- How can I connect to visual aids such as pictures, tables, and graphs?

- How can I use concrete materials, illustrations, and demonstrations to enhance mathematical learning?

- What words/pictures/representations are important to write on the board to develop a mathematical storyline of the entire lesson?

- What am I going to ask students to write during the lesson? Which examples of student work should the class discuss?

*Source:* Questions adapted from Chval, Chávez, Pomerenke, and Reams (2009).

Teachers have the opportunity to use specific teaching strategies that enhance multilingual learners' experiences and help them fully participate in mathematics classrooms. Throughout each of the following chapters, you will continue to delve deeper into strategies that will facilitate the participation of multilingual learners and position them as leaders in your classroom.

## Reflect

- How will you identify factors that hinder participation for multilingual learners in your mathematics classroom?

- What strategies will you use to enhance participation for multilingual learners?

# CHAPTER 4
# FACILITATE PARTNERSHIPS BETWEEN MULTILINGUAL LEARNERS AND THEIR PEERS

## Key Concepts

In this chapter, you will

- ✓ consider the importance of developing productive peer partnerships for multilingual learners.

- ✓ identify instructional strategies to facilitate these productive partnerships in mathematics classrooms.

- ✓ plan how to apply these principles in your classroom practice.

At one point while participating in the three-year-long professional development intervention, Ms. Bristow—whose third-grade mathematics classroom we visited in Chapter 2—began to observe student behavior in partnerships that concerned her. Specifically, she witnessed situations where partners dominated interactions with multilingual learners Julia and Mateo. For example, one student grabbed Julia's pencil out of her hand to complete her mathematics work for her. In another situation, a partner positioned the handouts in front of himself and away from Mateo, who thus was positioned as a spectator. In these interactions, Julia and Mateo had limited opportunities to communicate due to their partner's domination and, at times, were ignored or disrespected. As a result, Ms. Bristow began to question and reflect on her decision-making practices related to partnerships. The interactions Ms. Bristow observed and reflected on between her multilingual learners and their peers are not unique to her mathematics classroom, but transcend all classrooms in all contexts because they reflect human behavior, drives, and values.

## SOLVING A MATHEMATICS PROBLEM WITH A PARTNER

Imagine you are a fifth-grade student and you and your partner have been given the following problems to solve. How would you approach this situation with your partner?

1. 수진이네 축구팀에 8명의 선수가 있습니다. 선수들은 연습을 마치고 서로 하이파이브를 합니다. 각 선수끼리 서로 한 번씩 하이파이브를 한다면, 모두 몇 번을 하게 됩니까?

*Source:* iStock.com/pijama61

2. 유나는 서점에서 16,000원짜리 책을 사려고 합니다. 1,000원, 5,000원, 10,000원 지폐를 이용하여 16,000원을 지불하는 방법은 모두 몇 가지입니까?

3. 다음 직사각형의 넓이는 얼마입니까?

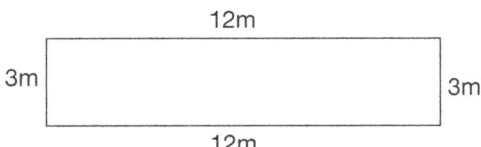

(We thank Sheunghyun Yeo for providing us with these problems.)

You and your partner would likely face challenges solving these problems because they are written in Korean. However, the images may have helped you try to reason. You may have also added the values in Problem 2, or you may have calculated the perimeter of the rectangle in Problem 3, although the text asked for the area. How would your partnership have changed if your partner read Korean?

## WHAT THE RESEARCH SAYS ABOUT PEER–TO–PEER INTERACTIONS WITH MULTILINGUAL LEARNERS

Multilingual learners' academic growth can be fostered through

- collaboration with multilingual peers who are fluent in the same first language, which will provide social and emotional support as they navigate learning in contexts taught in English;

- the delegation of authority to student groups, meaning students will control their work and discussions, while the teacher holds the students accountable for productive interactions; and

- interactions across languages or across language levels, such as interactions using multilingual learners' first language or dominant first language (e.g., going back and forth between languages, connecting and defining words) (Cohen & Lotan, 2004; Helfrich & Bean, 2011; Teale, 2009).

However, for this to occur in mathematics classrooms, teachers need to establish productive partnerships involving multilingual learners in and out of the classroom.

## QUALITIES OF PARTNERSHIPS

Facilitating effective partnerships in your mathematics classroom requires an understanding of qualities that illustrate productive and unproductive partnerships. Think about partnerships that you have seen that were productive, like the partnership shown in Figure 4.1. Think about those that have been unproductive. What were the qualities of each that made them productive or unproductive? Notate these qualities in Try It! 4.1.

**Figure 4.1** An Effective Partnership

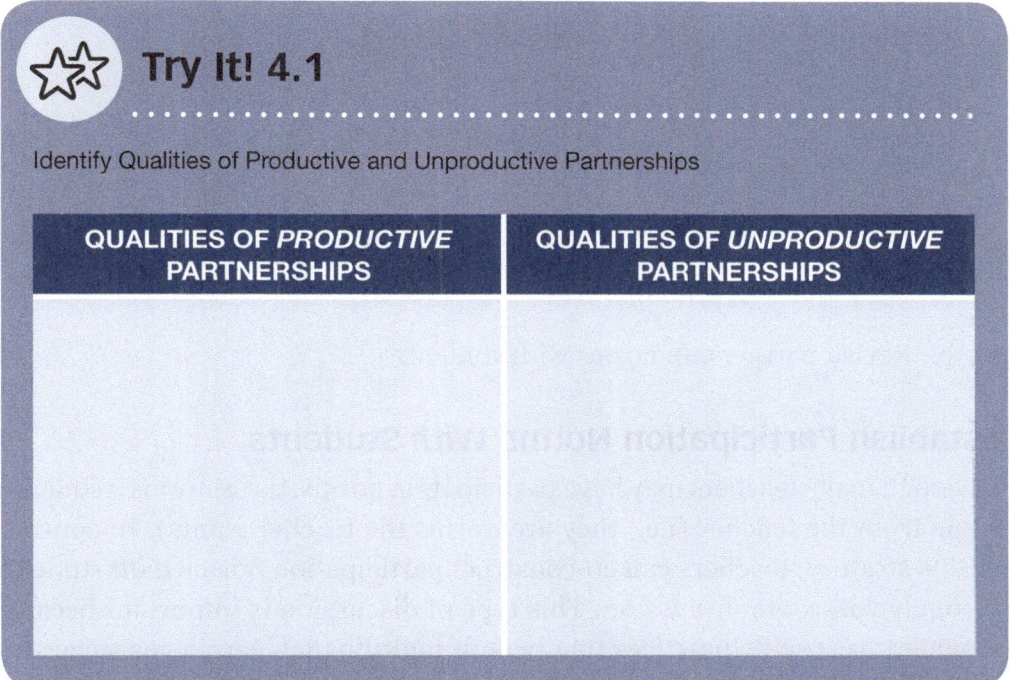

## Try It! 4.1

Identify Qualities of Productive and Unproductive Partnerships

| QUALITIES OF *PRODUCTIVE* PARTNERSHIPS | QUALITIES OF *UNPRODUCTIVE* PARTNERSHIPS |
| --- | --- |
|  |  |

Look across the two lists of qualities in Try It! 4.1. Notice there are a range of factors that may influence partnerships, such as students' strengths and mathematical competencies. In addition, establishing productive classroom norms (i.e., expected behavior agreed on by the members of the classroom) is critical to fostering productive partnerships in mathematics classrooms. In the remainder of this chapter, you will explore ways to establish classroom norms and identify different instructional strategies to foster productive partnerships for multilingual learners.

*Establishing productive classroom norms is critical to fostering productive partnerships in mathematics classrooms.*

## STRATEGIES FOR FACILITATING PARTNERSHIPS

In 2014, the National Council of Teachers of Mathematics (NCTM) established the Access and Equity Principle, which emphasizes the importance of the classroom community for student participation and engagement:

> Classroom environments that foster a sense of community that allows students to express their mathematical ideas—together with norms that expect students to communicate their mathematical thinking to their peers and teacher, both orally and in writing, using the language of mathematics—positively affect participation and engagement among all students (Horn, 2012; Webel, 2010). (NCTM, 2014, p. 66)

There are a number of strategies you can use to establish and maintain classroom norms that facilitate every student's participation:

1. Establish participation norms *with* students

2. Practice compliments

3. Provide concrete resources

4. Pay attention to pairing

5. Monitor how partners work

6. Revisit partnership norms with students

### Establish Participation Norms *With* Students

Although many teachers may have participation norms, these norms frequently come from the teacher (i.e., they are norms the teacher wants). In contrast to this strategy, teachers can co-construct participation norms *with* students through classroom discussion. This type of discussion is important because it generates specific practices that benefit multilingual learners as suggested by the NCTM (2014), such as additional wait time to process questions and compose verbal answers and the inclusion of all learners rather than privileging particular groups. Although you might initiate this discussion, you should draw on ideas from students since they are more likely to adhere to norms they generate themselves, such as

- listening and not interrupting when others speak,

- not speaking over a peer when they are talking,

- respectfully disagreeing with others' ideas,

- encouraging peers to participate, and

- using appropriate eye contact and body language.

Participation norms like these are also important because they can build the capacity of students, including students who may have been historically privileged in mathematics, to recognize and value the cultural and linguistic strengths of their multilingual peers and to work productively and equitably with peers who may come from different backgrounds than their own (Tabors, 2008). This is important as it prepares *every* student to work collaboratively in problem solving, a competency that will be required throughout their lives. For additional ideas about norms and characteristics of partnerships, see Figure 4.2. On the left-hand side, the ideas presented by Cobb (1995) will facilitate high-quality partnerships in mathematics classrooms. As you facilitate a discussion about possible norms, the ideas on the right-hand side may be introduced.

**Figure 4.2**  Norms and Characteristics of Productive Partnerships

| NORMS THAT SUPPORT PRODUCTIVE PARTNERSHIPS | CHARACTERISTICS OF PRODUCTIVE PARTNERSHIPS |
|---|---|
| • Persevere when solving challenging problems | ✓ Communicate and listen effectively |
| • Explain your solutions to a partner | ✓ Value other people and their ideas/strategies |
| • Listen to your partner's explanations and attempt to make sense of those explanations | ✓ Learn from and challenge one another |
| • Respectfully challenge explanations that seem unreasonable | ✓ Contribute to solving the assigned mathematical tasks and challenges |
| • Justify your thinking when challenged | ✓ Use time effectively |
| • Attempt to reach an agreement on a solution or solution method | ✓ Share the floor with others so there is an equitable distribution of roles, duties, and responsibilities so that every student is positioned for success |

*Source:* Reprinted with permission from *Access and equity: Promoting high-quality mathematics in grades 3-5*, copyright 2018, by the National Council of Teachers of Mathematics. All rights reserved.

## Practice Compliments

The norm of mutual respect is critical to any mathematics classroom if students are to productively engage in mathematical discussions that challenge their thinking. One way to foster mutual respect—a basis for productive partnerships—is using a routine Ms. Bristow referred to as "compliments" in her third-grade classroom. (This routine was in addition to the bookmark students had taped to their desk that they could consult while working together, as described in the next section.) For this routine, Ms. Bristow introduced an activity where students could write compliments to one another during their spare time and slip them into a compliment basket (see Figure 4.3). Ms. Bristow read compliments that were written by students (anonymously) at the end of the day as they prepared to pack up and leave for the day. Of course, Ms. Bristow slid a few compliments into the basket to ensure that compliments were written about all the children in her classroom. To illustrate this practice, let's peek

**Figure 4.3** Basket of Compliments

inside her classroom when Ms. Bristow was found reading some of these compliments. All the students are sitting on the carpet at the front of the room. Ms. Bristow sits in a chair with the basket filled with compliments on her lap. She then begins by pulling out one compliment and reading it to the class, directing her attention to Kim, the student who is receiving the compliment. Then Ms. Bristow reads a second compliment about Gina:

> 66 *The first compliment is to Kim. "Kim lets me play with her every time. She uses her manners. She is an awesome friend. She shares and takes turns and never makes fun of me, either. She never complains about a game that I want to play. Kim rocks." Good job. To Gina, "You played with me when no one else would. You rock." Good job.* 99

Although the routine of reading compliments fell outside of class time devoted to mathematics, it reinforced the norm of mutual respect across the entire school day. In situations where teachers may only see students for one class period a day, this routine could still be used with slight modifications. For instance, teachers might collect compliments across a week and read them on Fridays. These compliments could come from interactions both in and outside of the mathematics classroom.

## Provide Concrete Resources

For multilingual learners in particular, and students with verbal processing challenges, providing concrete resources can support students as they enact classroom norms and make language connections (Chval, Pinnow, Smith, & Rojas Perez, 2018). For example, Ms. Bristow used a bookmark (see Figure 4.4) that was taped to each student's desk and could be consulted during partner work. Ms. Bristow decided to include sample questions and compliments on her bookmark. This provided students with explicit things they could say in peer conversations while focusing the discussion on mathematical thinking. Bookmarks are one way to provide students with tangible aids to facilitate peer-to-peer conversations while doing mathematics together.

**Figure 4.4**  Text on Ms. Bristow's Bookmark to Facilitate Conversations

**Questions**

- What did you do first?
- How can we start?
- What do you think?
- How did you solve that problem?
- Would drawing a picture help?
- How can I help you?
- How else could we solve this problem?

**Compliments**

- I like your math thinking.
- I like your picture.
- Thank you for helping me.
- You did a good job at . . .

*Source:* Reprinted with permission from *Access and equity: Promoting high-quality mathematics in grades 3-5,* copyright 2018, by the National Council of Teachers of Mathematics. All rights reserved.

## Pay Attention to Pairing

When partnering students, you should consider not only the characteristics of productive partnerships and classroom norms, but also partner characteristics and compatibilities. Look back at Try It! 4.1. What are some of the qualities you listed for productive and unproductive partnerships? How many of these qualities are individual characteristics?

## STOP AND THINK

Stop and think about the strategies you use or have used to partner students.

- What individual student characteristics or compatibilities do you think about when you create partnerships for mathematical work?

When you partner students, you may consider communication style, personality, relatability, demeanor, participation, supportiveness, encouragement, patience, and/or willingness to yield control. It is also important to consider the cultural values of your multilingual learners since cultural values shape how students engage with the teacher and peers. For instance, students from cultures that place high value on group cohesion and group harmony (e.g., China, Japan, Korea) or on social character and a respectful attitude toward the teacher (e.g., Mexico, Central America) will privilege those values in their interactions with others even when it means they may be dominated by peers from cultures that encourage assertiveness or competitiveness (Pinnow & Chval, 2015). This can set up dynamics in the classroom that require the teacher to observe and address conflicts in values that do not support productive partnerships. Regardless of which criteria you use, it's important to consider students and their individual contexts, including personalities and character traits, mathematical competencies, and language competencies. Now, imagine you were going to have two new students join your class tomorrow. What qualities would you look for when selecting partners for each? Try this out in Try It! 4.2.

 **Try It! 4.2**

Choose Partners for Multilingual Learners

Pat and Sam are two new students who will join your class tomorrow. Here is some information about each of them.

> Pat is a third-grade student who has just joined your class after recently moving to Missouri from New Mexico, where Pat's family lived for two years. Pat's family relocated to be closer to extended family and possible new financial opportunities as the family's financial situation was bleak in New Mexico. Pat is the middle child of five, who all live with their mother. Since moving to Missouri, Pat has not adjusted well socially and has been unable to make friends. Pat is very quiet in class, appears timid, and does not participate in class discussions.
>
> Sam is also a third-grade student who has just joined your class after moving to Missouri from Mexico. Sam's family relocated for new job opportunities. In Mexico, Sam's family was financially stable and had the privilege to travel annually. Sam's family is composed of a younger sister, Nancy, and both parents. Since moving to Missouri, Sam has had trouble integrating into the school community and making friends. In class, Sam has not performed well academically and has picked a fight with another student.

- What qualities would you look for when selecting partners for Pat and Sam?

- Why are these qualities important for each of these specific students?

Now, imagine you received additional information on Pat and Sam—in particular, that both are multilingual learners with different proficiency levels in English. Pat has a high proficiency in English due to attending a dual language program in New Mexico. Sam has had limited opportunities to learn English.

## STOP AND THINK

Stop and think about this new information.

- What additional qualities would you look for in selecting partners for Pat and Sam during a mathematics lesson?

- Why are these additional qualities important?

(We thank Óscar Rojas Pérez for creating these descriptions of Pat and Sam.)

Sometimes teachers may only think about a multilingual learner's language proficiency when considering a productive partner and background other individual qualities. However, it is important to consider characteristics beyond just language proficiency when pairing students since language competencies are being revealed and developed *through* peer-to-peer interactions. Just like you do for other students, you should consider each individual multilingual learner's characteristics, competencies, and compatibilities when creating partnerships.

*It is important to consider characteristics beyond just language proficiency when pairing students since language competencies are being revealed and developed through peer-to-peer interactions.*

The strategies you use to partner students may change over the course of the academic year. For instance, at the start of the year you may be more selective with partnerships as students develop tools for navigating challenging interactions (Pinnow & Chval, 2015). Still, it is important to keep in mind that the ultimate goal is that every student can productively interact in partnerships with any other student in class by the end of the year.

## Monitor How Partners Work

There are subtle cues that teachers can learn to notice and address when observing students' partnership activities. Important cues include the following:

- *Who speaks first* when the teacher joins the group?

- *Which* student speaks for the group when the teacher approaches the group or interacts with the group?

- *How* did this student become the spokesperson for the group (i.e., was this a group decision, or did one student simply assume that role)?

  – Are the same students the spokesperson during partnering work?

  – How often are multilingual learners the spokesperson for their group, and how did this status evolve in the group?

▶ *Who controls the materials* that have been used for problem solving? When partners use manipulatives, *where are they located* on the table when the teacher joins a particular group? Are the materials regularly located in front of one particular student and not others? *Do students appear to share materials*, with multilingual learners having access to those materials? *Who uses these materials* when explaining what the group did?

▶ What happens *when multilingual learners speak or share their ideas*? Do the other students listen respectfully, or do they speak or make noises while multilingual learners are speaking? Do other students interrupt or talk over multilingual learners so that they have difficulty completing their ideas aloud?

▶ *What do patterns of disagreement look like* in these partnerships? Do all students know how to disagree without being disagreeable? For instance, once a multilingual learner has shared an idea, do other students consider the idea and respectfully question or disagree, or does the interaction begin to devolve into unproductive interactions? (Chval et al., 2018, pp. 124–125)

As you monitor how partners work and use these questions to reflect on your observations, there are a number of different steps you can take in response. For instance, you can check in with the partners, remind them of your classroom norms, encourage them to reference their bookmarks, model appropriate interactions or behaviors, or task them with developing their own strategies to address their unproductive partnership. To prepare you for this work, imagine you encountered the six situations in Try It! 4.3 with different pairs of students, one a multilingual learner and one a native English speaker, during a mathematics lesson (Chval et al., 2018, p. 120). Given these situations, how would you respond as a teacher?

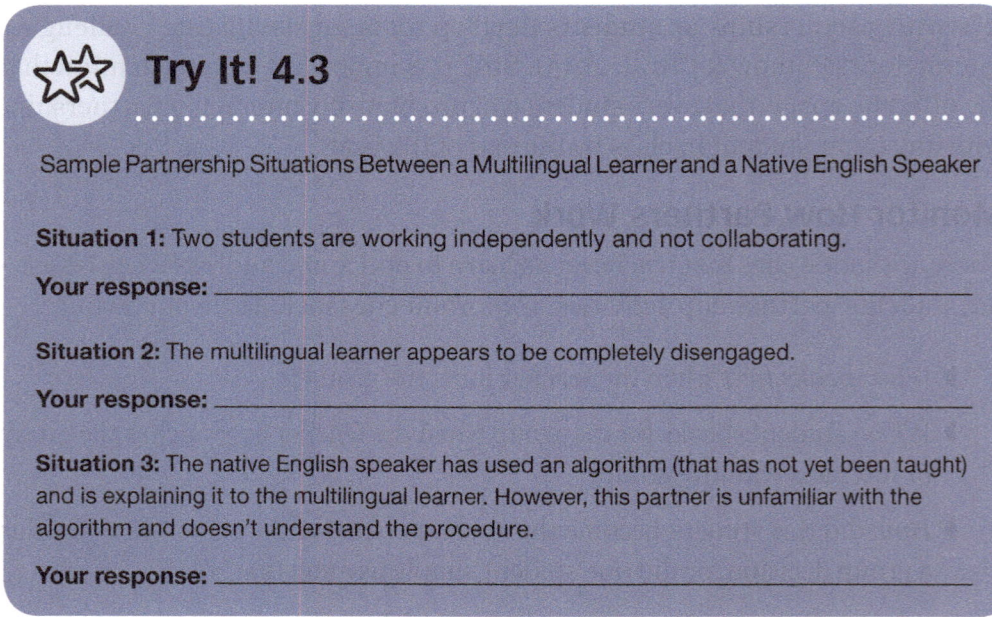

### ⭐ Try It! 4.3

Sample Partnership Situations Between a Multilingual Learner and a Native English Speaker

**Situation 1:** Two students are working independently and not collaborating.

Your response: _____

**Situation 2:** The multilingual learner appears to be completely disengaged.

Your response: _____

**Situation 3:** The native English speaker has used an algorithm (that has not yet been taught) and is explaining it to the multilingual learner. However, this partner is unfamiliar with the algorithm and doesn't understand the procedure.

Your response: _____

**Situation 4:** The native English speaker is completing all the work for both students while the multilingual learner spectates.

Your response: _____

**Situation 5:** The native English speaker grabs the pencil from the multilingual learner and begins to write on the multilingual learner's paper.

Your response: _____

**Situation 6:** The native English speaker is continually dismissing the multilingual learner's thinking and strategies.

Your response: _____

As you considered each of these situations, you may have planned to intervene with a specific partnership. However, sometimes this is insufficient and requires a broader conversation with the class where partnership norms are revisited.

## Revisit Partnership Norms With Students

Students may not always act in ways that demonstrate an understanding of productive partnerships. However, these instances can be opportunities to revisit partnership norms with students. We worked with Ms. Bristow for three years. The word *partner* was first spoken by Dr. Chval in the second planning session. Following that planning session, Dr. Chval selected a video filmed by a multilingual learner to discuss with Ms. Bristow in their first debrief session. After watching that clip, Ms. Bristow initiated a conversation about partners. Ms. Bristow recognized that facilitating productive partnerships required continued work throughout the year and could be challenging. In a later conversation, Ms. Bristow realized that the strategy she used in her literacy lessons of explicitly revisiting partnership norms could easily be applied during her mathematics lessons. In a subsequent mathematics lesson, we captured this discussion and present it in Transcript 4.1. This is one of multiple lessons we captured where Ms. Bristow was seen revisiting partnership norms with her students.

In this lesson, Ms. Bristow noticed some partnerships faced challenges and chose to spend the end of class discussing some of her observations. In this closing discussion, Ms. Bristow first role played with a student on how body language can convey messages of disrespect to partners. Then, Ms. Bristow reiterated qualities of productive partners and how students can use their bookmarks (Figure 4.4) to be quality partners. Let's look at how this played out.

**Transcript 4.1**

| | |
|---|---|
| Ms. Bristow: | Usually we talk about math at the end, but we are going to talk about partnerships because there were some partners who had conflicts today. What is a resource for you if you are a partner in math? What can you look at if you are like, "Hmm . . . I do not know what to say to this person?" Leland? |
| Leland: | You can look on your math partner sheet [Figure 4.4]. |
| Ms. Bristow: | Yeah, we have bookmarks that have questions, and they even have compliments. There will be times where it will be hard to find something nice to say to somebody. They may be driving you a little crazy, but this can help you. You may not want to say anything at all, but there is a whole list of compliments that you can say to somebody—"I like your math thinking" or "Great job" or "Thank you for helping me." And other things are a compliment to them, like smiling at them or being nice to them. I have seen a lot of this. I have not had a lot of partners say, "No, I do not want to work with you," but I have had some other situations. Like if Mike is my partner and I am sitting with Mike and Mike is trying to work with me and I am sitting like this [*arms crossed, looking away*] . . . am I saying anything rude to Mike? |
| Alex: | No, but you are acting rude. |
| Ms. Bristow: | What do you mean acting rude? |
| Alex: | Like you are like not even paying attention. |
| Ms. Bristow: | Okay, so is my body language respectful? |
| Toni: | No! |
| Ms. Bristow: | No, and your body language is just as important as what you are saying. If my body language is like this and Mike is working with me, how might he be feeling? What is Mike thinking if I am acting like this, Carol? |
| Carol: | He will not want to be your partner anymore because you are not paying attention. |
| Ms. Bristow: | Yeah. Are you always going to be a partner with your best friend? |
| Students: | No. |
| Ms. Bristow: | No. Sometimes you are not going to be. My best friend in my class was never my partner because we chatted all of the time, and so my teacher never allowed Kelly and me to be partners because all we wanted to do was talk and I had to get over it. I had to be able to meet new people and be kind. And it is not always what you are saying—your actions can be hurtful. So, if you are stomping—or grouchy when you have that person selected as your partner—that is not being a kind partner. Okay? So, look at your bookmarks [Figure 4.4]. Make sure you are asking people questions—make sure that you are giving people compliments because that is what we need in partnerships. Okay? So please remember that body language is just as important as the words that you are saying to each other. Does that make sense? Thumbs up or thumbs down . . . Got it? Okay, so tomorrow I will be looking for you guys to be showing respectful body language with your partner by nodding when they are talking to you and by smiling and being kind. |

In Transcript 4.1, you may have noticed how Ms. Bristow capitalized on an unproductive partnership she observed during the lesson to reinforce desirable behaviors of mathematical partners and highlight how body language can relay messages. By taking the time to discuss qualities of productive partners during a mathematics lesson, Ms. Bristow reiterated the importance of partnership to students' mathematical learning.

## THINKING ABOUT PARTNERSHIPS IN YOUR PRACTICE

In this chapter, you looked into the classroom of Ms. Bristow to examine student partnerships. By examining her practice, you saw how Ms. Bristow established classroom norms and facilitated productive partnerships between multilingual learners and their peers when she noticed inequities in partnerships. Use the ideas presented in this chapter as well as the questions that follow to reflect, initiate discussions with colleagues, and develop plans to enhance partnering situations in your classroom and school. Try It! 4.4 asks you to fill in the three columns of a table with your ideas for how to facilitate productive partnerships with multilingual learners before, during, and after a mathematics lesson. As you plan and think through lessons with a focus on multilingual learners, you will enhance your teaching practice while contributing to multilingual learners' mathematical success.

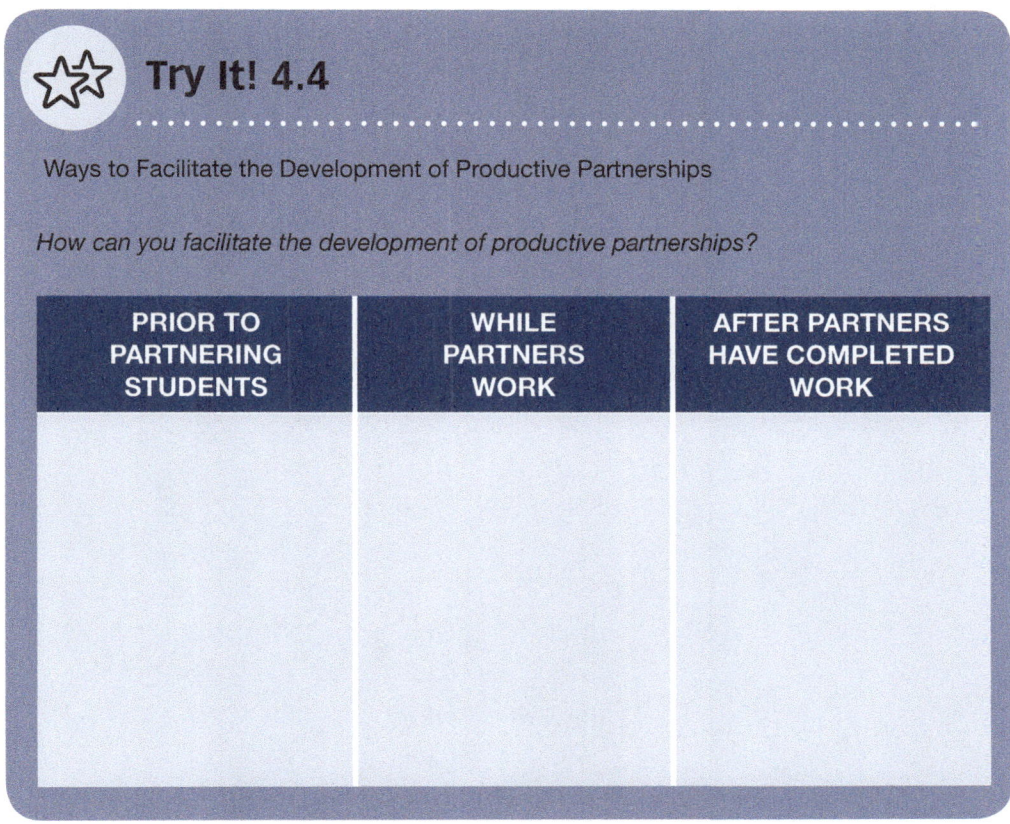

**Try It! 4.4**

Ways to Facilitate the Development of Productive Partnerships

*How can you facilitate the development of productive partnerships?*

| PRIOR TO PARTNERING STUDENTS | WHILE PARTNERS WORK | AFTER PARTNERS HAVE COMPLETED WORK |
|---|---|---|
| | | |

## Reflect

- How do I establish an environment where students respect one another and value partnerships?

- What criteria should I use when selecting partnerships for multilingual learners, and how does this change over the course of the school year?

- What strategies should I use to handle situations involving unproductive partnerships with multilingual learners?

# CHAPTER 5
## ENGAGE MULTILINGUAL LEARNERS THROUGH CULTURALLY RELEVANT CONTEXTS

## Key Concepts

In this chapter, you will

- ✓ understand why meaningful contexts are important to multilingual learners' mathematical learning.

- ✓ identify which contexts are relevant and meaningful for your multilingual learners.

- ✓ create a lesson rooted in a culturally relevant context for your multilingual learners.

As teachers, we have often come across mathematical problems that are situated in curricular contexts unfamiliar to our students and even ourselves. In other cases, we have seen contexts that students are familiar with (e.g., a grocery store), yet the actions embedded in the context are unlikely or undesirable (e.g., buying 100 watermelons). Over time, our repeated exposure to ineffective contexts began to raise questions: When does the context interfere with students' mathematical learning? How much instructional time do I invest in building the meaning for this context? How should I select contexts to use with *my* students? In this chapter, you will explore some answers to these questions.

## REFLECTING ON THE CHALLENGE OF CONTEXTS AND CULTURE

Imagine you are a fourth-grade student and have been given the following problem to solve. How would you approach this problem?

> In a T20 match Australia bats first and scores 261 runs. John and Sam open for England. John scores 3 runs off every ball he faces while Sam scores 2 runs off every ball bowled to him. Australia never bowls an illegitimate ball or gives any extra runs. Assuming John faces the first ball and Sam would face the last two balls of the match, how many balls will it take for England to win the match?
>
> (We thank Chintan Mehta for providing this problem.)

It is likely you are facing challenges with this problem and need more information on the context.

### STOP AND THINK

Stop and think about your feelings when you first read the problem.

- How might your feelings mirror the feelings that multilingual learners experience as they solve problems involving unfamiliar contexts?

Here is some additional information about the game of cricket from Wikipedia. It may help you solve the problem.

Feeling better about how to solve the problem?

> **"** *The score of a cricket team whose innings is in progress is given as the number of runs they have scored "for" the number of wickets their opponents have taken. For example, a team that has scored 100 runs and lost three wickets has a score of "one hundred for three," written 100–3 (also 100/3); [t]he exception to this is in Australia, where it is conventional to reverse the wickets and runs scored, so that what would be*

*written 100–3 elsewhere in the world is written and said 3–100 (or 3/100) in Australia. A team that is dismissed having scored 300 runs is said to have a score of "three hundred all out," rather than "three hundred for ten"; the score for the innings is then simply written 300. However, if a team declares their innings closed (in a First-class match) or reach an over limit (in a limited-overs match), the number of wickets is included in their score for the innings, for example 275–7. A declaration is noted by appending a "d" or "dec" to the score (for example 300–8d); such a score is spoken in the standard form with the word "declared" appended (example: "300 for 8, declared").*

*In a two-innings match, the scores of each team for their two innings are always given separately rather than being totalled, and the current score is stated in terms of how many runs the batting team lead or trail by. An example of a score for a two-innings match in progress would*

How confident are you feeling now?

*be: Team A 240 & 300–7d, Team B 225 & 130–4. This indicates that Team A in their first innings scored 240 runs, and Team B made 225 in reply. Team A then made 300 for 7 in their second innings, declaring it closed, and Team B are currently 130 for 4—in this scenario, Team B currently trail by 185 runs (and are said to be facing a target or run chase of 316, the score with which they would win the match) . . . .*

*When a game is completed, there are standard ways of referring to the difference in scores between the two teams. For instance, if Team A, batting first, scored 254–6, then Team B, batting second, only scored 185, whether or not they go all out, it would be said that "Team A won by 69 runs" because they either bowled out the opposition or caused them to exhaust their overs (in a limited-overs match) when they were trailing by 69 runs. On the other hand, if team A, batting first, scored 254–6 but team B, batting second, scored 255–8, it would be said that "team B won by 2 wickets" because they reached their target with 2 wickets remaining. In a two-innings match, a team can win having only batted once (while the other team, batting twice, has not equaled the other team's score). For instance, team A score 160 all out, team B score 530 and declare, then team A score 230 all out. In this case it would be said that "team B won by an innings and 140 runs." 〞*

*Source:* Scoring (cricket). (2020, June 28). In *Wikipedia.* https://en.wikipedia.org/wiki/Scoring_(cricket)

## STOP AND THINK

Stop and think about your feelings before and after you received more information about the context.

- Have your feelings changed? Why or why not?

As you likely recognized from your experience trying to solve the cricket problem, contexts that students have little familiarity with, or knowledge of, can create unintentional barriers that restrict students' access to mathematics. As a teacher, it's important to recognize that even though you may use district-adopted curriculum materials, they may not include contexts that are relevant or meaningful to all of your students. Instead, the contexts may be more reflective of the curriculum writers' interests or experiences than your students' prior knowledge, experiences, or interests. When selecting contexts to use with your multilingual learners, it's critical they are culturally relevant and connect with students' own experiences (as opposed to your experiences). As Bay-Williams and Livers (2009) state,

*When selecting contexts to use with your multilingual learners, it's critical they are culturally relevant and connect with students' own experiences (as opposed to your experiences).*

Cultural relevancy means that students can relate to the topic and become engaged in the problem because they understand its context and find it interesting. Culturally relevant contexts are not always obvious. Consider, for example, a group of multinational students in an urban setting who are exploring fractions in the context of farming. Will the farmer's plan to devote portions of his land to various crops serve as a relevant context for these students to learn about fractions? Students may have experiences in farming and understand the concept of wanting to select different crops to ensure profitable harvest, in which case the problem *is* a culturally relevant example. However, this context may have little interest for students or may be completely unfamiliar, in which case the context will not be serving its purpose of supporting student thinking about the mathematical ideas. (p. 241)

Think about how your interest, or lack thereof, in cricket affected how willing you were to engage in learning more about the mathematics problem. As a teacher, it's important for you to modify your curriculum to incorporate contexts in which your students can connect (we will see more about adapting curriculum materials in Chapter 12). Yet, this does not mean you should only use contexts your students already know because this will do little to prepare them for situations when they encounter problems with unfamiliar contexts. Instead, you must be strategic in selecting contexts that can expand their understanding of the world.

You have likely used a range of different contexts to engage and motivate your students while simultaneously illustrating the relevance of and connection to mathematics in their daily lives. You may have also used contexts to provide students with opportunities to draw on informal strategies they might use in the given situation, such as divvying out objects equally.

## STOP AND THINK

Stop and think about the contexts you have used.

- What are examples of real-life and familiar contexts you have used in your mathematics classroom?

- When were these contexts successful? When were they unsuccessful? What contexts were familiar to your multilingual learners? Which were unfamiliar?

- When can contexts or the introduction of contexts distract multilingual learners from learning mathematics?

## WHAT THE RESEARCH SAYS ABOUT CONTEXTS IN MATHEMATICS

Mathematical contexts that are related or connected to multilingual learners' life experiences can support their language and mathematical learning (Barwell, 2003; Domínguez, 2011; Domínguez, López Leiva, & Khisty, 2014; Secada & De la Cruz, 1996); however, there is not one curriculum that is relevant to all students (Gutstein, 2003). As a result, teachers need to adapt or create curriculum materials that are relevant to their students. It is not enough to simply insert a real-life context into a problem for the context may not be meaningful or familiar to students. Thus, teachers need to carefully consider the mathematical contexts they use and build meaning for them if students are unfamiliar. One way to identify relevant and meaningful contexts is to draw on students' community and cultural knowledge (Chval, 2010; Vomvoridi-Ivanovic, 2012; Wager, 2012). As teachers create or adapt curriculum materials, it is important that the task does not become simplified, but maintains its mathematical rigor (Chval & Chávez, 2011). The goal is to *amplify* curriculum, not just seek to simplify it (Gibbons, 2015).

> *Teachers need to carefully consider the mathematical contexts they use and build meaning for them if students are unfamiliar.*

## STRATEGIES FOR ENGAGING MULTILINGUAL LEARNERS THROUGH CULTURALLY RELEVANT CONTEXTS

Meaningful contexts can serve as rich resources for creating mathematics tasks or problems. In addition, they can be used to advance multilingual learners' mathematics and language learning, and to share aspects of multilingual learners' cultures. In recent years, teachers have increasingly used videos in mathematics lessons to establish meaningful contexts. You can use videos you find online or can easily create them in your community using your smartphone, which can be shown during lessons with technology (e.g., a projector, an interactive whiteboard). In addition to using videos from the United States, you can use videos from countries that will honor multilingual students in your classroom (e.g., a marketplace in Turkey for students from that country).

Ms. Bristow wanted to design lessons that were effective for all of her students. She sometimes decided to use videos to introduce meaningful contexts in her lessons. As she selected videos, she considered her multilingual learners. In addition, Ms. Bristow considered other students when selecting videos. During two of the years we filmed in her classroom, she had a student who was unable to hear and another who was unable to see. Thus, she thought about other accessibility features such as subtitles and speed of speech (words per minute) in the video.

To illustrate her use of video in her third-grade classroom, you will read two transcripts from her lessons and two descriptions of videos she used. Across these examples, you will see how Ms. Bristow used three different contexts that were strategically selected. As you examine the three scenarios, notate the following for *each*:

- The meaningful and attractive elements of the context for students

- The potential of the context for creating mathematics tasks/problems

- The strategies Ms. Bristow used to unpack the meaning of specific language

- The time required to establish a shared understanding of the context

### Context 5.1

Oftentimes teachers use store contexts in mathematics lessons. To introduce problems associated with money, Ms. Bristow could have used a store context in a major city two hours away with which some students in her class would be familiar. However, Ms. Bristow recognized that approach would create an obstacle for some of her students. Instead, Ms. Bristow selected a store that was closest to the school, increasing the likelihood that all the students, including multilingual learners, had either shopped there or seen it on the way to school. Let's take a peek at how this played out in the lesson (see Transcript 5.1). To begin the lesson, Ms. Bristow showed her class a brief two-minute video that Dr. Chval filmed at a local dollar store near the school. In the video, Dr. Chval is shown walking up to the dollar store; entering it; walking throughout; and picking up a range of items, including a stuffed bunny rabbit, a small soccer ball, a crossword book, and a box of candy. Then, Dr. Chval is shown at the register purchasing the items that totaled $7.42 with cash and coins.

## Transcript 5.1

| | |
|---|---|
| Ms. Bristow: | Remember how yesterday we were talking about ways to make a dollar? |
| Class: | Yeah. |
| Ms. Bristow: | You guys came up with a lot of different ways to make a dollar yesterday. Has anyone ever been to a store like Dollar Box? Where most things are about a dollar? |
| Class: | [*Many hands raised*] Yeah. |
| Ms. Bristow: | Most things are a dollar, but then there's always some things that are a little bit more or a little bit less. Have you guys been to one of those stores before? |
| Class: | Yeah. |
| Alex: | There's a Dollar Box where they have some really neat stuff. |
| Ms. Bristow: | They have some cool stuff . . . . Today we are actually going to go on a little video field trip to the Dollar Box. Actually, Dr. Chval went to the Dollar Box, and she's the one that went for us to show us what's going on at the Dollar Box. Now we are going to take a little field trip to Dollar Box. Alright, so find a spot that you can see the video [*plays video*]. She [Dr. Chval] had a chance to go to the Dollar Box. How many friends have been to the Dollar Box or a store like that? |
| Class: | [*Hands raised*] |
| Ms. Bristow: | I love going to the Dollar Box. I always feel like I'm getting a good deal. When we were in the Dollar Box, you saw a lot of signs for things that are like a dollar, but do you remember how they had at the cash register some candy, those Sugar Babies? Those weren't a dollar—they were actually less than a dollar, right? And so, they [Dollar Box] can't print up all price tags for a dollar because sometimes they want to charge a different price. What I'm going to have you guys do first thing today is to make some price tags for some different items. So, you're going to have to show me how you're going to write a certain price [*shows on the board a box of candy and stuffed rabbit with blank price tags hanging on them*]. What is a price tag? Do you guys know what a price tag is? . . . So there's a price tag on a couple of these items that Dr. Chval bought [*holds up bag of items bought at Dollar Box*]. On this rabbit [*holds up stuffed rabbit*] there is a price tag. On this plush squeeze ball [*holds up ball*] there's a price tag. What is a price tag? What does that mean, "price tag"? Samuel, what's it mean? |
| Samuel: | It means how much it costs. |
| Ms. Bristow: | Yeah, so a price tag tells you how much something costs. And I did a couple pictures here. So usually there is some kind of little tag. Like on this little bunny there's a tag [*holds bunny up by its tag*]. It would tell you how much it costs, okay? So, we're going to make some price tags today. |
| Dani: | Is a price tag the one with a whole bunch of lines on it? |
| Ms. Bristow: | That one's a bar code, and that's what tells the person at the register how much something costs [*writes on board "Price tag: Tells you how much something costs"*] . . . . So, a price tag tells you how much something costs. I have some different items that were in the video for the Dollar Box [*projects image of worksheet with different pictures of items on it*]. I've written out the price of items in words. So, the Sugar Babies were sixty-nine cents. There was some applesauce, and I said that was ninety-seven cents. This checkers game was two dollars and eighty-nine cents, and this soccer ball, I think we have it, this was one dollar even [*holds up ball*]. So, what I'm going to have you guys do is write some price tags because this manager at the Dollar Box, he needs to label these items, right? I mean, when you're shopping, don't you want to know how much something costs? It helps me make a choice if I'm going to buy one thing or another. So, I need to know how much it costs. So, you're going to make some price tags today for just four different items. I want to see how you can write this in money form. So, I'm not going to do an example because I want to see how you guys do it, and then we're going to come back and share how we wrote this. |

## STOP AND THINK

Stop and think about Context 5.1.

- What were the meaningful and attractive elements of this context for students?

- What was the potential of this context for creating mathematics tasks/problems?

- What strategies did Ms. Bristow use to unpack the meaning of specific language for this context?

### Context 5.2

To provide a context for students to reason about arrays, Ms. Bristow introduced her class to a local business in her community called The Candy Factory. Since some of the students in her class had not had the opportunity to visit the business, Ms. Bristow used a video to introduce this unfamiliar context. This business has the store on the first floor and the factory where the candy is made on the second floor. In the video, Dr. Chval visits The Candy Factory in a way that captures students' interests—like the Dollar Box visit. In the video, Dr. Chval enters the business, looks through the glass display at all the different kinds of candy, and then asks to purchase a box of 10 chocolates. After the clerk gathers the chocolates, she shows them to Dr. Chval, who sees they are arranged in a two-by-five array. Next, Dr. Chval goes upstairs to the factory where customers can see the staff making different kinds of candy, such as dipping apples into caramel and filling molds with chocolate, and organizing them into arrays. After Ms. Bristow showed the video, she had a discussion with her students about the ways the chocolates, apples, and molds were organized. In this discussion, she introduced the term *arrays*.

## STOP AND THINK

Stop and think about Context 5.2.

- What were the meaningful and attractive elements of this context?

- What was the potential of this context for creating mathematics tasks/problems?

### Context 5.3

While teaching a unit on place value, Ms. Bristow read the story "Grandma Eudora's T-Shirt Factory" (Fosnot, 2007) to the class. Although some students had parents who worked in a factory, the majority of students had never actually been inside one. To provide students with the background knowledge of what a factory is, Ms. Bristow showed a video clip of a factory that creates something the children use every day—crayons (see Transcript 5.2).

**Transcript 5.2**

| | |
|---|---|
| Ms. Bristow: | We've been talking about a T-shirt factory, and I wasn't sure if you guys had ever been to a factory or seen a factory. So, I wanted to show you what a factory looked like before we move on with our T-shirt factory, and I found a video of the Crayola® crayon factory. They don't always let people inside the crayon factory, but this reporter got to go and look inside the factory and see how crayons are made and see how the boxes of crayons that you guys get at the beginning of school are made. So, I was really wanting to show you guys this video. So, I'd like you guys to watch this and see what steps they use for the crayons, and then we're going to compare the steps that they use to make the crayons to the steps that we're going to use for the T-shirts. Okay, so watch closely [*shows video*]. |
| Ms. Bristow: | [*Pauses video where open boxes full of crayons are shown and are on wood pallets*] Hold on, do you see in the video all these boxes of crayons? What do you think those are? What do you think they have all those boxes of crayons for? Darrel, what do you think the box is for? |
| Darrel: | A warehouse, the box is for . . . |
| Ms. Bristow: | The box is what? |
| Darrel: | What the box is for? |
| Ms. Bristow: | What are all these boxes for? What do they have them there for? What are they doing there? |
| Toni: | They're making crayons . . . making the crayons in boxes . . . making them there. |
| Ms. Bristow: | And store them there, right? So, "they're ready because all those boxes are stored in that warehouse" is what you [*points to Darrel*] said it was, right? |
| Darrel: | Yeah. |
| Ms. Bristow: | Okay, so they're storing those boxes in the warehouse, and they're waiting for what? What are they waiting for, Yasmin? |
| Yasmin: | For a truck to take them to put them in the stores. |
| Ms. Bristow: | Yeah, so they're waiting for people to order the crayons, and then they're going to be put in a truck and go to a store. |

## STOP AND THINK

Stop and think about Context 5.3.

* What were the meaningful and attractive elements of this context?

* What was the potential of this context for creating mathematics tasks/problems?

**VIDEO 5.1:**

Freedawn Home. (2015, October 30). *Inside the Crayola® factory where 12m crayons are made every day* [Video]. YouTube. https://www.youtube.com/watch?v=fK3-su7aQ5w

The clip Ms. Bristow references in Transcript 5.2 is no longer available. However, there are similar videos online. Watch Video 5.1, "Inside the Crayola® Factory Where 12m Crayons Are Made Every Day" (Freedawn Home, 2015); Video 5.2, "Inside the Crayola® Factory: See How the Iconic Crayons Are Made" (NBC Universal, 2017); or Video 5.3, "How People Make Crayons" (The Fred Rogers Company, 1981) from *Mister Rogers' Neighborhood*.

You may have noticed across the three contexts that Ms. Bristow did each of the following:

**VIDEO 5.2:**

NBC Universal. (2017, March 9). *Inside the Crayola® factory: See how the iconic crayons are made* [Video]. *Today.* https://www .today.com/video/ inside-the-crayola- factory-see-how-the- iconic-crayons-are- made-893853251852

▶ Integrated meaningful and relevant elements based on her students' interests and experiences. For instance, the items purchased at the Dollar Box were items her students would likely buy themselves.

▶ Chose contexts with which students were familiar and unfamiliar. For those that were unfamiliar to students, Ms. Bristow used this as a way to expand students' understanding of different contexts while advancing their language development across a wide array of topics and experiences.

▶ Used contexts that could be used across multiple mathematical concepts. For example, Ms. Bristow could circle back to the Dollar Box or Crayola® crayon factory when focusing on future mathematical concepts. This can save time since you don't have to build meaning for a brand-new context and instead you build on the class's shared experience from earlier lessons in the year.

**VIDEO 5.3:**

The Fred Rogers Company. (1981). *How people make crayons* [Video]. *Mister Rogers' Neighborhood.* https:// www.misterrogers.org/ articles/factory_visits/

▶ Selected contexts where mathematical concepts were reflected in the actual experience depicted, such as boxing up chocolates in arrays, and to share aspects of multilingual learners' culture (e.g., marketplace, social gatherings, money). Oftentimes contexts are selected that don't match the actual experience of learners, such as finding the number of people and pets at a park given a total number of legs (Wager, 2012).

▶ Highlighted and built meaning for language from the contexts that students would encounter in the curriculum materials, such as *warehouse, boxes, factory,* and *shipping.*

▶ Created videos and pictures on a smartphone or handheld camera or pulled them from the internet. Ms. Bristow did not spend a lot of time or money creating videos.

Using culturally relevant contexts in mathematics teaching not only provides rich resources to build mathematics and language, but also positions students' "out-of-school experiences as resources for learning rather than deficits to overcome" (Turner & Celedón-Pattichis, 2011, p. 151).

## STOP AND THINK

Stop and think about some contexts that are culturally relevant to your multilingual learners.

• What are some of your initial ideas?

## THINKING ABOUT CONTEXTS IN YOUR PRACTICE

In this chapter, you looked into Ms. Bristow's classroom and examined how she selected and built meaning for mathematical contexts for multilingual learners. In addition, you reflected on your own selection and use of contexts within mathematics. What would this look like in your classroom? Develop a lesson or series of lessons rooted in a culturally relevant context in Try It! 5.1.

 **Try It! 5.1**

Select a context connected to your multilingual learners' life experiences that can be used to motivate the mathematics of the lesson. When you select contexts in this way, students will be encouraged to draw on informal strategies used in the context to solve the problems (Wager, 2012). Moreover, such problems provide an opportunity to connect school mathematics with reality, and vice versa. If you're unsure what kinds of contexts would be appropriate, talk with your multilingual learners and their families to learn about their cultural practices and experiences and immerse yourself in their communities. After you have selected a context, use the following questions to guide your lesson construction.

1. What mathematics content would you address using this context?

2. What are specific words or terms that are involved in this context? How familiar are your students with the language involved in the context? What are some words and terms whose meanings need to be unpacked?

3. What strategies would you use to unpack the meaning of the language you identified?

Create a mathematics task for your students using the context you selected and taking into account the mathematics content and language you identified.

*If you're unsure what kinds of contexts would be appropriate, talk with your multilingual learners and their families to learn about their cultural practices and experiences and immerse yourself in their communities.*

## Reflect

- What do I need to consider when selecting contexts for multilingual learners?

- How can I use real-life and familiar contexts to advance multilingual learners' mathematical and language learning?

- What strategies will I use to build meaning for the contexts I use to teach mathematics?

# CHAPTER 6
## REACH MULTILINGUAL LEARNERS WITH VISUALS AND GESTURES

### Key Concepts

In this chapter, you will

✓ explore the importance of the use of visuals and gestures as instructional strategies with multilingual learners.

✓ identify how visuals and gestures can support multilingual learners as they learn mathematical concepts.

We often hear the recommendation to use visuals and gestures to enhance teaching, especially for multilingual learners, who may rely more heavily on multimodal forms of communication (like visuals and gestures) to make sense of academic content delivered in class. You have likely observed the benefits of using photographs, videos, objects, illustrations, drawings, manipulatives, representations, and physical models in teaching all learners. You have probably reinforced your spoken words by writing them on a visible surface so multilingual learners could compare the sound with the written text. You have likely used your hands to communicate or to emphasize meaning while explaining a concept (e.g., when you wanted students to be quiet or stop in their tracks). There is no question that the use of visuals will support learning. The old expression, "A picture paints a thousand words," comes to mind. Take, for instance, the photo from Ms. Bristow's third-grade classroom in Figure 6.1.

> *There is no question that the use of visuals will support learning.*

▶ What do you notice in the photo?

▶ What do you anticipate students will likely do and learn about mathematics in this task?

**Figure 6.1**  Making Trail Mix in Third Grade

You may notice a variety of measuring tools, bowls with different ingredients, and multiple sheets with directions. In this lesson, Ms. Bristow was using a recipe for trail mix to teach fractions.

While Ms. Bristow's use of visuals, objects, and tools in the lesson involving trail mix was engaging and effective, we have noticed instances where visuals and gestures could have been used more strategically. As we worked with teachers, we began to ask: How can we use visuals and gestures to facilitate learning for multilingual learners in mathematics classrooms? When are they helpful? Not helpful?

## WHAT THE RESEARCH SAYS ABOUT VISUALS AND GESTURES

The term *visuals* is actually a bit tricky because many people often consider a visual to be a nonverbal representation of an idea. Yet, a visual can represent graphical information that includes written language (Danielsson, 2016). Sometimes teachers use visuals with multilingual learners as they think a visual will lessen the linguistic load the student must bear in meaning making. Sometimes this is true, but we actually do *read* all visuals since learning to read a visual accurately requires particular competencies, just as reading written language does. Also, some visuals may oversimplify important concepts, making them less useful in teaching than one might hope. However, if used effectively in conjunction with other instructional tools, visuals and gestures can support multilingual learners as they acquire both language and mathematical knowledge (Domínguez, 2005; Hill & Flynn, 2006; Kersaint, Thompson, & Petkova, 2014; Lemke, 2003; Shein, 2012). These representations can help students learn, clarify, and reinforce mathematical language and concepts (Avalos, Medina, & Secada, 2015). Furthermore, the incorporation of visuals into curriculum materials can aid student comprehension of abstract concepts while providing a connection to language (Miller & Warren, 2014; Raborn, 1995). According to Radford (2009),

> Thinking, hence, does not occur solely in the head but in and through language, body and tools. As a result, and from this perspective, gestures, as a type of bodily action, are not considered as a kind of window that illuminates the events occurring in a "black box"—they are not clues for interpreting mental states. They are rather genuine constituents of thinking. (p. 113)

Visuals provide a unique support that can enhance students' mathematical learning through a drawn depiction of mathematical concepts (Avalos et al., 2015; Chval, Chávez, Pomerenke, & Reams, 2009). Moreover, the incorporation of gestures can support instruction through the reinforcement of verbal messages and concepts, emphasize meaning, and provide students access to communicate in nonverbal forms (Alibali, Nathan, & Fujimori, 2011). In Try It! 6.1, challenge yourself to explain two mathematical concepts without the use of visual or gestural aids.

## Try It! 6.1

Imagine you have to explain two concepts, *square* and *angle*, in your mathematics class without using visuals, gestures, or the words *square* and *angle* in your explanations. In addition, you cannot use your hands or any resource other than your voice. What will you say? How will you describe concepts such as width, base, acute angle, edge, column, and polygon? Try it with friends, colleagues, or students. Ask them to listen to your explanation and draw on a piece of paper, without using words, an image of what they understand from the explanation.

- What was challenging for you as you described the concept?

- What was challenging for those who drew what you described?

## USING VISUALS STRATEGICALLY

As you think about using visuals strategically, you likely ask yourself questions such as these: Should I use photos, drawings, or videos to support multilingual learners in this lesson? When should I draw a picture? When should I ask students to draw representations? When should I invest time to bring in physical models from home? These are challenging and important questions. Of course, the answers depend on many variables (e.g., time, new or unfamiliar language or concepts, students' mathematical and language proficiencies). Let's examine some examples from Ms. Bristow's third-grade classroom.

### Examining Ms. Bristow's Use of Visuals: Dividing Cookies

Ms. Bristow introduces the mathematical concept of division by reading the children's book *The Doorbell Rang* (Hutchins, 1986). As she reads, she shows the pictures from the book and poses questions to her students. After she finishes reading, she displays 12 cookies and a plate on the SMART Board® and asks the children how many cookies she would get if she shared the cookies with 1 person. She also writes "12 cookies shared with 1 person" on the board, then asks, "How can I show that 12 cookies are shared with 1 person?" After students respond, Ms. Bristow moves all 12 cookies onto the plate. She then introduces and writes the representation, $12 \div 1 = 12$, as shown in Figure 6.2.

**Figure 6.2**  Ms. Bristow's SMART Board®: 12 Cookies Shared With 1 Person

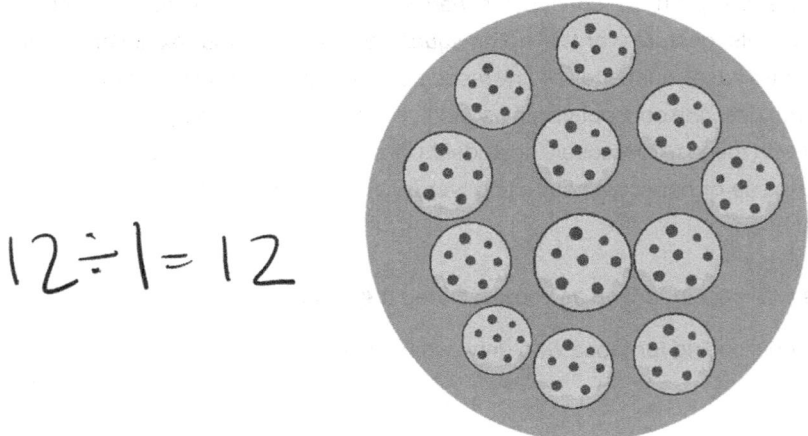

## 12 cookies shared with 1 person

$$12 \div 1 = 12$$

After Ms. Bristow finishes dividing the 12 cookies by 1, she asks the students how she should divide the 12 cookies shared with 2 people. One student says she has to split them up. Ms. Bristow explains that to organize the cookies on the SMART Board®, she needs some plates. She asks students how many plates she needs to share the cookies with another person. Ms. Bristow then displays two plates on the SMART Board®. She splits the cookies on the plates, but puts 11 cookies on her plate and 1 cookie on the other plate (see Figure 6.3). She then asks students if they agree with that way of sharing cookies. They disagree!

**Figure 6.3**  That's Not Fair

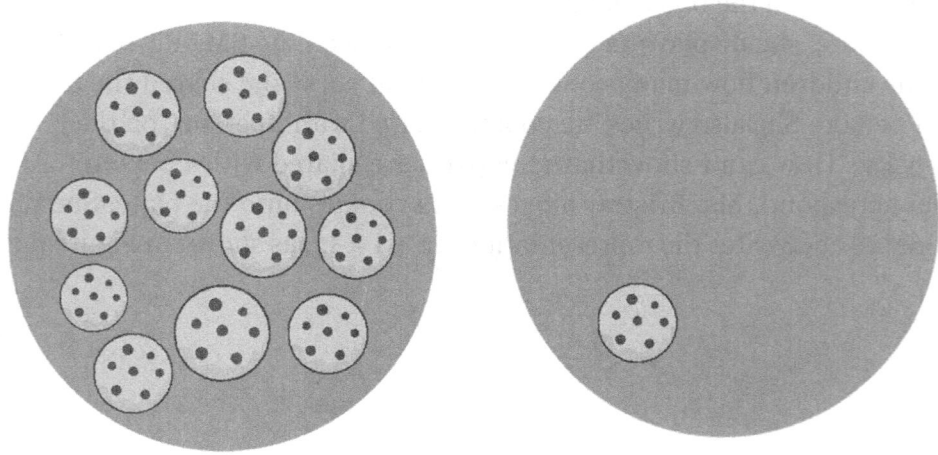

## 12 cookies shared with 2 people

Then, she asks what the problem is if she is sharing. One student says it is not fair to take more cookies than her partner. Then Ms. Bristow writes the word *fair* on the SMART Board® and asks, "What does 'that something is fair' mean?" Students respond with words such as *even*, *equal*, and *same amount*. Then she asks the students how she can share the cookies with another person fairly. She emphasizes that in division situations, fair shares are necessary. After Ms. Bristow does a few more examples of sharing 12 cookies with 3, 4, and 6 people, she distributes paper cookies and plates for students to solve fair-share situations at their desks. Therefore, students see visuals on the SMART Board®, but also have opportunities to manipulate physical models as they begin to solve division situations. Further, this activity mirrors the story and connects to real-life situations students likely have experienced.

## STOP AND THINK

Stop and think about the mathematical concepts you introduce to your multilingual learners.

- Which mathematical concepts are challenging to introduce and could confuse multilingual learners?

- How can you use visuals to facilitate multilingual learners' understanding of mathematical concepts that you teach?

## Examining Ms. Bristow's Use of Visuals: Emphasizing the Meaning of Loose

Earlier in the school year, Ms. Bristow used a curriculum unit focused on place value by Catherine Twomey Fosnot (2007). She began the unit by reading a story, "Grandma Eudora's T-Shirt Factory." In this story, a family creates rolls of T-shirts in amounts of 10. In addition to using the images in the book along with gestures, Ms. Bristow used physical models (i.e., some of her T-shirts from home) to demonstrate a roll of 10 T-shirts. In planning the lesson with the researcher, Ms. Bristow recognized that she would need to emphasize the meaning of loose shirts in this context, as the word *loose* had confused some of her students in a previous lesson. Read Figure 6.4 to examine the planning of the lesson (left-hand column) and then what transpired during the lesson (right-hand column).

**Figure 6.4**    Emphasizing the Meaning of *Loose*

| PLANNING EXCERPTS | | LESSON EXCERPTS | |
|---|---|---|---|
| Researcher: | Okay, and another word would be *loose*. That means it's not part of something else—it's not attached to something else. | Ms. Bristow: | We have this other word that we talked about in our read-aloud today. We've got *lose* and *loose*. What does *lose* mean? What does *lose* mean, Elijah? |
| Ms. Bristow: | Okay. | Elijah: | Like "lose a race." |
| Researcher: | I would also write on the board *lose* and *loose* and write both words underneath each other and [ask students,] "What do you notice about how these two words are spelled? And what's the different meaning of these words?" Say you want to get at "What does this word mean? How are they [*lose* and *loose*] different?" | Ms. Bristow: | "Lose a race"—is that what you said? Lose money. Okay, it could mean that you don't win something like you didn't win your basketball game, or it could mean something else. What else does *lose* mean? Grace, what else does it mean? |
| | | Grace: | When you lose a race. |
| | | Ms. Bristow: | That's what Elijah said—when you don't win something—but there's another type of *lose*. Amber? |
| | | Amber: | If you lose something that you like. |
| | | Ms. Bristow: | What do you mean "you lose something that you like"? |
| | | Amber: | Like if you lose a teddy bear and you, um, couldn't find it after school or something. |
| | | Ms. Bristow: | So, maybe you have something and you can't find it anymore [*writes on the board "you can't find something"*]. These words look awfully similar [*points to the words* loose *and* lose *on the board*]. We've got *lose*, and then this is *loose*, so what does *loose* mean? Because they are different words. What does that mean, Greg? |
| | | Greg: | Like you have a rabbit in a cage and then he got out, and he's running around the house and you can't find him. |
| | | Ms. Bristow: | Okay, so you've got something in a container. You've got something in a cage or a package, and then it gets out? |
| | | Sam: | Yeah, and it's running around your house and you can't find it. |

*Source:* Chval, K. B., Pinnow, R. J., & Thomas, A. (2015). *Mathematics Education Research Journal.* Used with permission.

## STOP AND THINK

Stop and think about how Ms. Bristow discusses the distinction between the terms *lose* and *loose*.

- Why would that be useful for multilingual learners?

## Emphasizing Mathematical Connections

Let's pick up with the next portion of the planning session and lesson in Figure 6.5.

**Figure 6.5**    Connecting the T-Shirts to Mathematical Models

| PLANNING EXCERPTS | LESSON EXCERPTS |
|---|---|
| **Researcher:** One thing you could do is model the roll. Like you take the 10 shirts, and here's a loose shirt. Another way we could say this is [with] a single shirt. I think you want to give some other words to it. You're going to use *loose*, but I think you need for them to build meaning for *loose*. Because that's going to be an unusual word for them to make sense of. | **Judy:** Like in the story they rolled up 10 T-shirts, and so they're all loose and some didn't make 10. |
| **Ms. Bristow:** That was something I was going to ask you about. I thought today after we rolled up the T-shirts to figure out what representation we're using of the rolls. I mean, what place value block are we representing the roll [with]? That was a confusing thing for Miguel, and well, I mean, it was confusing for several kids. I think this group of girls that was working here also had some trouble with that as well. | **Ms. Bristow:** Okay, so she's saying maybe like there were some that were in the rolls and there were some that were by themselves, maybe? So, there were, like, some that were a part of something not in a group [*writes on board*], so let me show you. Britney? What do you need, dear? What do you think? |
| **Researcher:** Right. So, I think what I would do is we've got the—we have a picture of the T-shirts. Let's say then we have a picture of the blocks. So, you need a picture that represents rolls. So, . . . a rubber band or whatever you decide you want to use. It could just be a single stick if I think about it. With the blocks, what would it be? | **Britney:** I've got another one like when a T-shirt's too loose. |
| | **Ms. Bristow:** Oh, you mean like when something's kind of baggy [*types on SMART Board® "something is baggy"*]? So what kind of loose are we talking about? Which one? Something that's baggy? A T-shirt that's baggy? Something that's in a cage and gets out or something that's not in a group? Emilio? |
| | **Emilio:** [*Provides an example of 13 with 3 not in the roll*] |
| | **Ms. Bristow:** Something that is left out is what I hear you saying. So, here's a roll, and this is worth—this is worth how much? |
| | **Students:** 10. |
| | **Ms. Bristow:** Okay, well then, what about the place value blocks? Which one matches the roll? Which one matches the roll? Britney, can you grab the one that matches the roll? |
| | **Britney:** [*Hands Ms. Bristow the blocks that represent the roll*] |
| | **Ms. Bristow:** [*Takes roll and holds a place value rod of 10*] Okay, so this is a roll, and this is worth how much? |
| | **Students:** 10. |

*Source:* Chval, K. B., Pinnow, R. J., & Thomas, A. (2015). *Mathematics Education Research Journal.* Used with permission.

During the planning meeting, words such as *lose* and *loose* were discussed to help Ms. Bristow bring out the nuances of these words in this curricular context. Two strategies were discussed in the planning meeting: (1) the use of visual images to help multilingual learners make links to the language being discussed and (2) the joint construction with students of the multiple and varied definitions of these words. The discussion regarding the multiple definitions of certain words was also cast against a broader cultural context by taking into account that linguistically and culturally diverse students often bring significant experiences to bear on their approach to school curriculum that are not always represented in the curriculum itself.

## STOP AND THINK

Stop and think about your teaching practice.

- What are ways you connect physical objects or written visuals to mathematical manipulatives or mathematical representations so that multilingual learners make connections?

## REFLECTING ON YOUR USE OF GESTURES

Some people purposely select and use gestures to help them communicate specific messages or meanings while in other situations they may move their hands in expressive ways, but not for the specific purpose of conveying a message (i.e., "talking with their hands"). Gestures, a form of nonverbal communication, can convey specific messages or meanings.

## STOP AND THINK

Stop and think about gestures that you use.

- What gestures do you use in everyday life (e.g., when you say hello or want a student to stop or be quiet)?

- What gestures do you use to teach mathematics?

- What gestures do you use that may be unfamiliar to multilingual learners?

## USING GESTURES STRATEGICALLY

When selecting visuals or gestures to use with multilingual learners, it is critical to consider the way they will be used. For example, will the visual or gesture be used to connect to students' existing background knowledge and experiences? Will the visual or gesture be used to build a foundation for a new concept or idea? In the former case, familiar visuals or gestures may be used to build on background knowledge and experiences; however, in the latter case, unfamiliar

*When selecting visuals or gestures to use with multilingual learners, it is critical to consider the way they will be used.*

examples may be used to move students forward. As we observed mathematics teachers, we noticed different purposes and types of gestures. Let's take a look at a couple of examples from Ms. Bristow's third-grade class.

## Examining Ms. Bristow's Use of Gestures: Obstacle Course

Ms. Bristow planned an activity for the concept of elapsed time, a common concept for third grade. She based the activity on an upcoming field day at the end of the school year where the students would do different activities all day. The task prompted students to create their ideal field day. Students had to select activities and determine how much time they would spend at each activity. Ms. Bristow included a variety of activities on the handout, such as an obstacle course (see Figure 6.6).

**Figure 6.6**   Field Day Activities Handout

**Field Day Activities**

| Hula Hooping | Potato Sack Race |
| Castle Ball | Four-Square |
| Three-Legged Race | Kickball |
| Dance, Dance, Revolution | Water Play |
| Bounce House | Obstacle Course |
| Soccer | Tug of War |

**Directions**

1.  Choose from the activities listed above to make a schedule for Field Day.

2.  Decide how long you would like to participate in each activity.

3.  Write the start and end times in the chart below. *Note:* Field Day starts at 9:00 a.m. and ends at 12:00 p.m.

| Activity | Length of Time at the Activity | Start Time | End Time |
|---|---|---|---|
| Example: Soccer | 15 minutes | 9:00 a.m. | 9:15 a.m. |
|  |  |  |  |
|  |  |  |  |
|  |  |  |  |

*Note:* Ms. Bristow included 11 blank rows for her students.

When Ms. Bristow asked her students if they knew what an obstacle course was, one of her multilingual learners replied, "No." Ms. Bristow then began with a story about her young daughter visiting a local park that had an obstacle course. As she described her child's navigation of the obstacle course at the park, she used her hands and bent her body to describe her daughter's journey through the tires and the tunnels. All of Ms. Bristow's students loved to hear stories about her young daughter. Based on Ms. Bristow's effective use of gestures and her reference to a local park, her multilingual students knew the meaning of *obstacle course* before they began the activity. The multilingual learner who asked Ms. Bristow to explain an obstacle course replied with "I love those."

## STOP AND THINK

Stop and think about your use of gestures.

- What are examples of situations where you use or could use gestures to emphasize the meaning of language in your mathematics classroom?

Ms. Bristow's gestures involved more than pointing; they focused on the *meaning* of the term. Ms. Bristow could have used a picture or video to illustrate an obstacle course. Teachers must decide how much time to devote to the use of different media when they introduce contexts during their mathematics lessons. Sometimes, multilingual learners may demonstrate that they do not understand a specific term during a lesson, and the use of gestures may suffice in explaining it. As gestures convey meaning, they can support mathematics instruction because they reinforce verbal messages and concepts. Additionally, they help students communicate nonverbally (e.g., using thumbs-up). However, in other situations, a photo, drawing, or video may be necessary.

Examine the aims, types, and examples of gestures in Figure 6.7, adapted from concepts by Gullberg (2008, 2010), Kendon (2004), and O'Neill, Topolovec, and Stern-Cavalcante (2002), and think about the gestures you use to teach mathematics concepts. Try It! 6.2 asks you to fill in two rows based on examples you use in your mathematics classroom.

**Figure 6.7**   Aims, Types, and Examples of Gestures

| AIM | TYPE | EXAMPLE |
|---|---|---|
| To explain or describe concepts or actions | Descriptive | Using your hands and body to demonstrate how to move through an obstacle course |
| To point out something important or signal an element of importance | Referential | Tracing the perimeter of a rectangle with your finger to highlight a part of the problem |
| To specify the degree with relation to size, shape, distance, proportion, duration, speed, concentration, density, or rates of change | Explanatory | Using your hands to show the degree of a measurement such as size or distance (e.g., "The field is far away" or "Her shoe size is larger than mine") |

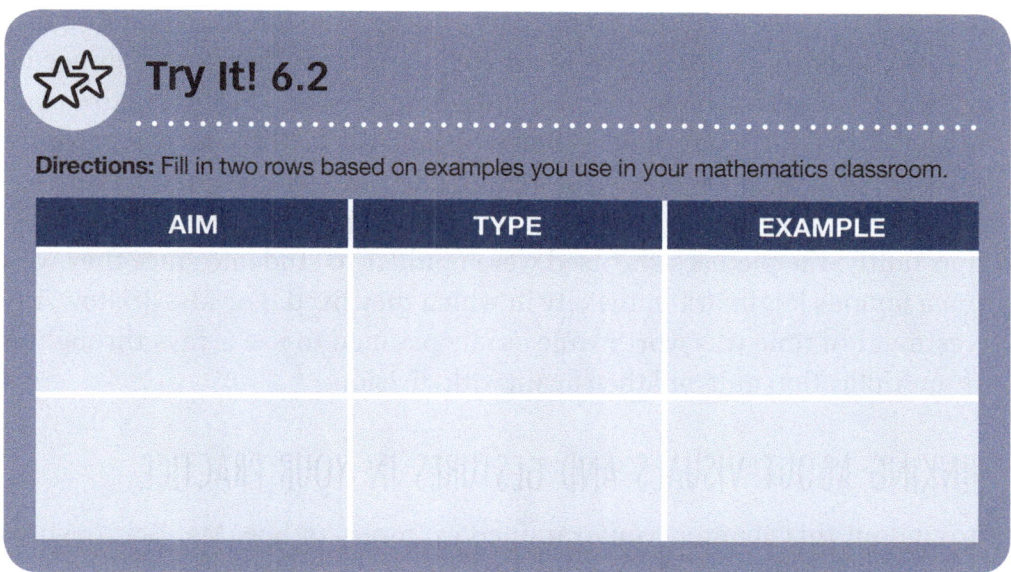

### Try It! 6.2

**Directions:** Fill in two rows based on examples you use in your mathematics classroom.

| AIM | TYPE | EXAMPLE |
|-----|------|---------|
|     |      |         |
|     |      |         |

## Examining Ms. Bristow's Strategic Use of Visuals and Gestures: Columns and Rows

Ms. Bristow decided to use arrays to develop meaning for the concept of area in her third-grade classroom. She recognized from experience that students often confuse the terms *columns* and *rows* when working with arrays. She displayed photos of columns in architecture and rows in an auditorium on her SMART Board®. She rearranged chairs in her room to discuss different arrays involving 24 chairs. In that case, she asked the children to identify which arrangement would be the best to watch a movie on the SMART Board®. The students did not select the 1 by 24 or the 24 by 1. In addition to the use of photos, Ms. Bristow used drawings, color, and gestures. Let's take a peek inside her classroom, via Transcript 6.1.

### Transcript 6.1

| | |
|---|---|
| Ms. Bristow: | So, you remember rows and columns because you know columns are like this [*holds arms vertically, to represent columns*], like the columns at X [*referencing one of the pictures on the SMART Board® of a famous local landmark*]. And rows are, like, rows—rows go across, so with our arrays [*holds arm horizontally, to represent rows; draws a rectangle with three columns and two rows*], which ones are columns, and which ones are rows? Yesenia? |
| Yesenia: | Which are the columns? |
| Ms. Bristow: | Which ones are columns? Which way do I do columns? |
| Yesenia: | Up and down. |
| Ms. Bristow: | Up and down, so this is a column, right [*draws a green arrow within one of the columns inside the rectangle on the SMART Board® and gestures up and down*]? |
| Yesenia: | Yes. |
| Ms. Bristow: | Okay, so I can write *column* [*writes the word* column *in green above the arrow*], and then which way is rows? Which way is rows, Ignacio? |
| Ignacio: | Side to side. |
| Ms. Bristow: | Okay, so this way is a row [*draws a purple arrow inside one of the rows in the rectangle, writes the word* row *in purple beside the row, and gestures across the row*]. |
| Ignacio: | Yeah. |

Ms. Bristow recognized that her students needed to be able to distinguish between 5 bundles of 3 things and 3 bundles of 5 things. Therefore, they needed to differentiate between rows and columns. In this example, Ms. Bristow made connections between the mathematical concept, the mathematical representation, and the language, spoken and written. She used photos, colors, drawings, gestures, and labels to help students develop meaning for columns and rows. Importantly, the pictures she used were familiar to students since they were from a famous landmark in the city in which they lived. For Ms. Bristow, this investment of time was worthwhile as she planned to use arrays throughout her multiplication unit and then again with division.

## THINKING ABOUT VISUALS AND GESTURES IN YOUR PRACTICE

Throughout this chapter, you examined examples of how Ms. Bristow used visuals and gestures, including physical objects, to help multilingual learners build meaning for columns and rows, as well as distinguish between *loose* and *lose*, which look similar but have different meanings and spellings. As you design your instruction, you will make decisions about the types of visuals and gestures to use for specific contexts and how much time to devote to their use. You will also consider how you will use visuals and gestures to connect language with mathematical concepts and support multilingual learners as they learn about the subtleties of the English language (e.g., the multiple meanings for *loose*, and the distinction between *loose* and *lose*).

*As you design your instruction, you will make decisions about the types of visuals and gestures to use for specific contexts and how much time to devote to their use.*

Figure 6.8 shows the mathematical work of Lucinda, a multilingual student in Ms. Bristow's class. In this lesson, Lucinda had the opportunity to work with physical dice, create drawings that represented the numbers on the dice, write mathematical expressions, and use gestures in her presentation as her class began to explore developing meaning for multiplication.

**Figure 6.8** Lucinda Using Different Representations on the SMART Board®

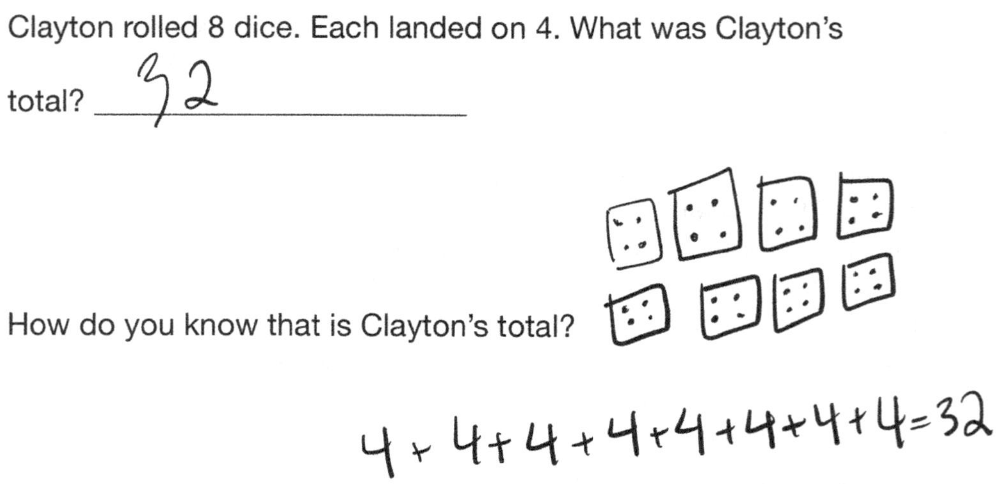

Clayton rolled 8 dice. Each landed on 4. What was Clayton's total? _____ 32 _____

How do you know that is Clayton's total?

$$4 + 4 + 4 + 4 + 4 + 4 + 4 + 4 = 32$$

It is important to note that as you discuss the different meanings of words, you may encounter situations when specific definitions are offensive in English or other languages (such as *loose*). Offensive meanings may also occur with some gestures when they have different meanings in different countries. In fact, some common gestures in the United States are offensive in other countries. If you have students from different parts of the world, knowing these distinctions may be important. You will find some blog posts online about this topic, such as the following:

- "6 Hand Gestures in Different Cultures (& What They Mean)" by Luciano Cipolla (2018)

- "15 Hand Gestures That Have Different Meanings Overseas" (2019)

- "5 Everyday Hand Gestures That Can Get You in Serious Trouble Outside the U.S." by David Anderson, Matthew Stuart, Mark Abadi, and Shayanne Gal (2019)

## STRATEGIES FOR USING VISUALS AND GESTURES

Read through the following list of research recommendations. Mark your current practices with a *C* and the practices you could emphasize more with an *M*. Then use Try It! 6.3 to plan for use of such practices in your mathematics classroom.

- I teach new language to multilingual learners by incorporating it into content rather than teaching it in isolation (Gibbons, 2015; Moore-Harris, 2005).

- I use stories with visuals to contextualize mathematical topics for multilingual learners (Gutierrez, 2002).

- I use visuals to connect new language and abstract concepts (Raborn, 1995; Takeuchi, 2015).

- I use gestures to emphasize important characteristics in mathematical representations so that multilingual learners can build meaning (Alibali et al., 2011).

- I use gestures to reinforce verbal messages and concepts, and help students to communicate nonverbally (Chval, 2004; Dominguez, 2005; Takeuchi, 2015). I encourage my students to do this as well.

**Try It! 6.3**

Identify a key mathematical concept or term that you will introduce in the next month and plan how you will use visuals and gestures to support student understanding. Discuss your decisions with a colleague.

## Reflect

- What are challenges associated with using visuals and gestures in your mathematics classroom?

- What should you consider when you make decisions about whether to use visuals and gestures? How to use them? When to use them?

- What would you tell a colleague are important considerations when using visuals and gestures to facilitate mathematical learning?

# CHAPTER 7
# ANALYZE MATHEMATICAL WORK OF MULTILINGUAL LEARNERS

## Key Concepts

In this chapter, you will

✓ understand the importance of analyzing multilingual learners' mathematical work to reflect and improve on your teaching practice.

✓ identify instructional strategies to facilitate the identification of multilingual learners' mathematical and language competencies.

Multilingual learners bring important experiences, competencies, and ideas to the classroom. Like any student, they can also face challenges in the classroom. When working with multilingual learners, it is important to recognize that there are many factors that can facilitate or hinder access to mathematics and not all are language related. For example, we have observed multilingual learners who demonstrated mathematical misconceptions with place value and fractions that were unrelated to language. Being able to recognize these differences is fundamental to creating opportunities for multilingual learners to learn. In our conversations with teachers, they have admitted that they have assumed a child could not solve a mathematics problem due to language, when in fact the multilingual learner was hindered by experiences with a specific mathematical concept. In this chapter, we provide opportunities for you to examine your own assumptions when examining multilingual learners' mathematical work.

Many teachers recognize that examining multilingual learners' work, in conjunction with observing these students engaging in mathematical tasks, can be helpful in determining their mathematical understandings. By analyzing and discussing specific examples of multilingual learners' mathematical activity with them, you can gain additional insight into the many ways students approach mathematical problems or develop specific concepts or proficiencies. When you complement observations and analysis of student work with mathematical conversations, a more robust picture of student thinking can be determined.

## REFLECTING ON YOUR EXPERIENCES

Think about a multilingual learner that you know or have had as a student. Write down the student's first name:

_____.

Imagine that no one examined this learner's work to assess mathematical thinking over the course of a year (i.e., only mathematical *answers* were examined to determine correctness). What is problematic about this situation?

### STOP AND THINK

Stop and think about when you observe multilingual learners solving mathematics problems and analyze their written work.

- What do you focus on in these situations?

## WHAT THE RESEARCH SAYS ABOUT ANALYZING MULTILINGUAL LEARNERS' MATHEMATICAL WORK

As multilingual learners are in the process of developing their language competencies, they may be unable to express their mathematical thinking in English at a given point in time. Therefore, teachers need effective methods of identifying multilingual learners' mathematical conceptions (Moschkovich, 1999). If teachers do not attend to the multiple demands of content knowledge and language acquisition, multilingual learners' mathematical learning will not develop to its full potential as academic content language plays a key role in understanding mathematical concepts (Gibbons, 2015; Khisty, 1995; Moschkovich, 2013, 2015; National Academies of Sciences, Engineering, and Medicine, 2018a). Moreover, it is important for teachers to understand that language use or familiarity does not equate to the use of academic language necessary to express mathematical thinking (Gutierrez, 2002). For example, individual students may have learned a mathematics concept or procedure in their first language in their country of origin. However, they may not be able to explain the solution in English based on opportunities to learn and use English in the context of mathematics. For multilingual learners to develop the mathematical language to effectively explain their thinking, they must be provided with opportunities to use mathematical discourse (Khisty, 1995; National Academies of Sciences, Engineering, and Medicine, 2018a). If such opportunities are not provided, the task of accurately identifying mathematical and language understandings or conceptions among students is especially difficult.

## EXAMINING JOAQUIN'S MATHEMATICAL WORK

Before you examine Joaquin's work, first try a similar task to the one he was provided. In Try It! 7.1, take time to read the instructions and examine the worked example (Figure 7.1). Then, fill in the shapes in Number Sentence 1 and Number Sentence 2.

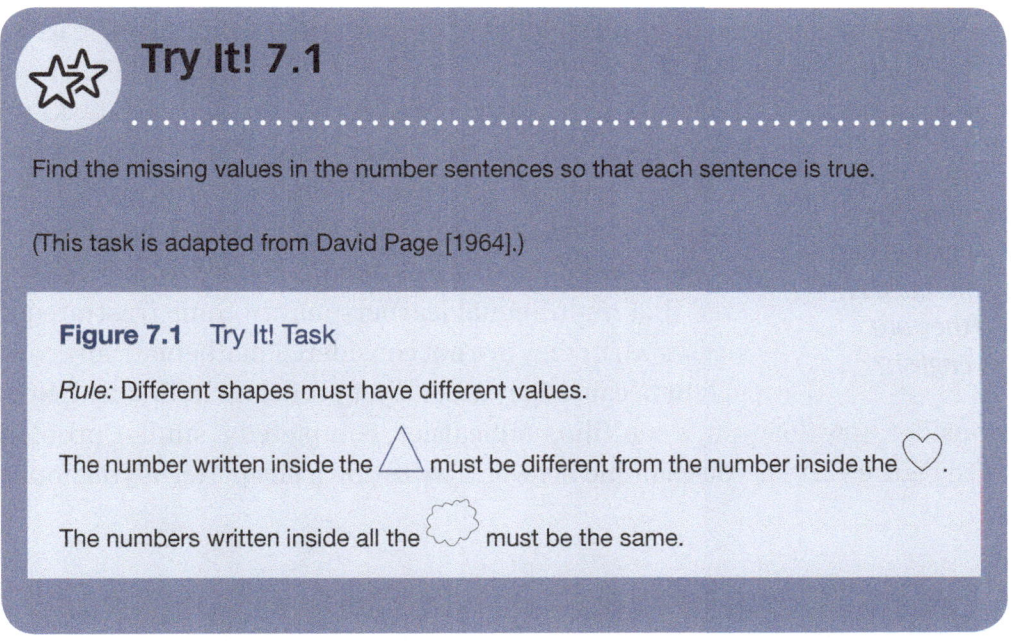

### Try It! 7.1

Find the missing values in the number sentences so that each sentence is true.

(This task is adapted from David Page [1964].)

**Figure 7.1**  Try It! Task

*Rule:* Different shapes must have different values.

The number written inside the △ must be different from the number inside the ♡.

The numbers written inside all the ☁ must be the same.

(continued)

(continued)

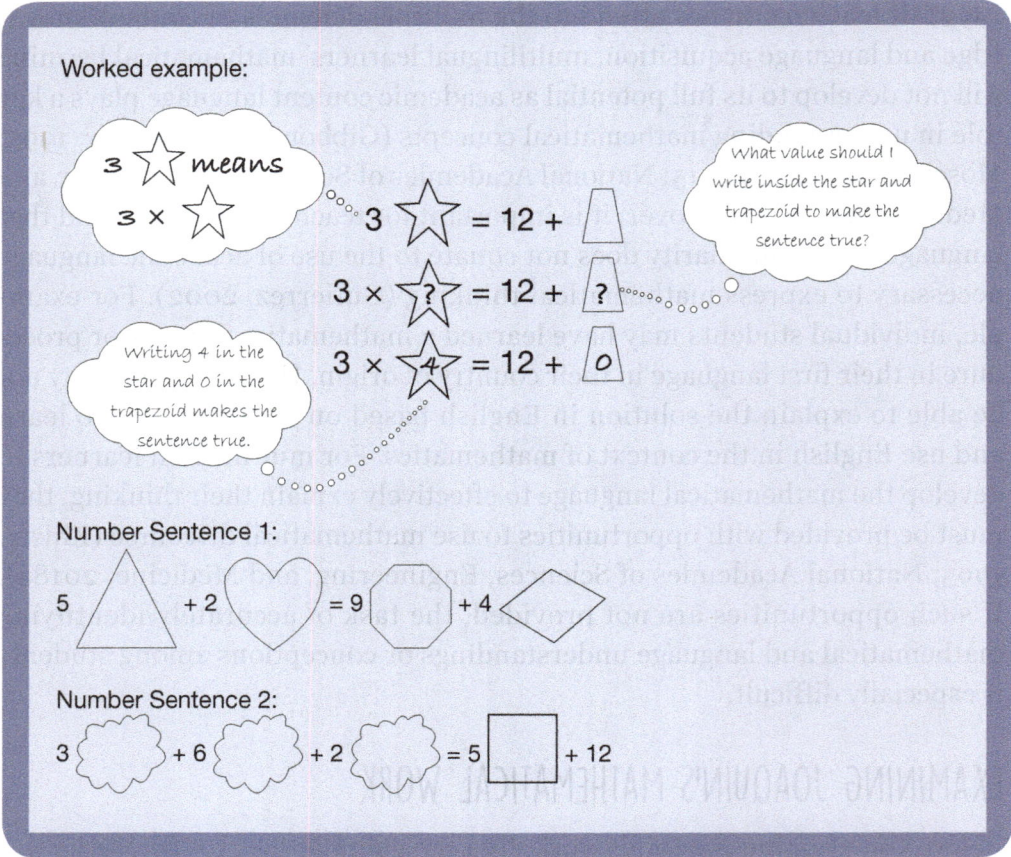

The problems posed in Try It! 7.1 offer multiple entry points for students, are cognitively challenging, and ensure access for a range of learners. Moreover, such problems encourage students' problem-solving skills and can facilitate students' transition to algebra. These kinds of problems can be used with a range of students and can be differentiated. Importantly, multilingual learners should not be excluded from challenging mathematical problems like Try It! 7.1. In our experiences, multilingual learners who are provided such tasks are highly engaged. It is not recommended that teachers wait to challenge multilingual learners until they master the English language at an advanced level since we learn language by using it (Gibbons, 2015). Have you ever considered that multilingual learners may become frustrated in classes when they are not considered mathematically competent because they are not yet proficient in English? Now, consider how Joaquin, a multilingual learner, completed a similar problem (see Figure 7.2). As you examine his work, write down any patterns you notice.

*Have you ever considered that multilingual learners may become frustrated in classes when they are not considered mathematically competent because they are not yet proficient in English?*

**Figure 7.2**   Joaquin's Work

Sentence 1:

$$2 \heartsuit = 8 + \diamondsuit$$

with 4 inside the heart and 1 inside the diamond

Sentence 2:

$$\triangle + \varhexagon = 10 + \bigcirc$$

with 3 in the triangle, 7 in the octagon, and 1 in the circle

Sentence 3:

$$10 - \boxed{3} = \boxed{7} + 1$$

with 3 in the quadrilateral, 7 in the square, and 1 in the cloud shape

As you examined Joaquin's work, you may have noticed all the things he understands. For instance, he solved the computation contained in each problem, which provides evidence that he understands multiplication, addition, and subtraction. Joaquin's work also illustrates an operational conception of the equals sign—a common student conception. In this conception, the equals sign means to "do something," so students assume the value immediately to the right of the equals sign represents the answer to the expression on the left-hand side. We can see evidence of this in Joaquin's work as he repeatedly ignores the value contained in the last shape in each sentence and populates each final shape (on the right-hand side of the equation) with 1. These two patterns illustrate Joaquin's conception of the equals sign. It is important to notice there is *no* evidence in Joaquin's work that indicates he is facing language challenges in understanding the directions. Thus, in this case, Joaquin's teacher would begin by examining his understanding of the concept of equivalence. You may want to see if your students hold a similar conception as Joaquin by asking them, "What does '=' mean?" Oftentimes children respond that it means to "get the answer"—hence, to "do something."

> *You may want to see if your students hold a similar conception as Joaquin by asking them, "What does '=' mean?" Oftentimes children respond that it means to "get the answer"—hence, to "do something."*

Joaquin's conception of the equals sign implies a view of the symbol as an operation and not a representative of equivalence (Behr, Erlwanger, & Nichols, 1976; McNeil & Alibali, 2005). When students encounter problems that challenge an operational conception, they may not know how to approach the problem or what to do (as Joaquin illustrates). Such conceptions are not unique or limited to elementary school students but can permeate through high school (Kieran, 1981) and restrict students' ability to be successful in algebra and beyond. Curriculum materials can perpetuate operational conceptions by fostering habits of mind among children through repeated exposure to specific patterns of use of the equals sign (McNeil et al., 2006). Importantly, research has found students who are successful at solving algebraic equations possess a relational, or equivalence, understanding of the equals sign (i.e., they understand the left and right sides of the equals sign are equivalent; Alibali, Knuth, Hattikudur, McNeil, & Stephens, 2007; Knuth, Stephens, McNeil, & Alibali, 2006). To develop a relational conception of the equals sign, students need explicit discussions and to experience a wide range of problems. It is important that students do not notice and develop such thinking on their own (Denmark, 1976). As a result, teachers need to explicitly discuss and foster relational conceptions of the equals sign beginning in the elementary years to foster students' mathematical success. In Joaquin's case, assistance from the teacher is required to shift his operational conception of the equals sign. As a result, teachers must be able to distinguish between different conceptions of the equals sign and consider ways to foster an equivalent perspective of it.

However, a looming question remains: How is this mathematical conception on the student's part related to the student's status as a multilingual learner? Don't other students have similar conceptions about equivalence? In reality, yes! However, the difference is that teachers can be quicker to evaluate a student's answer as an equivalence concept rather than as a language challenge when that student speaks English as their first or only language. Teacher assumptions play a key role in instances such as this one. If the teacher does assess a multilingual learner's answer as an equivalence issue rather than related to language, there is still the process of navigating this conception in an additional language for the multilingual learner as the student engages in mathematical questions, answers, and explanations. Addressing the conception in the student's additional language of English requires that the teacher ask probing questions, wait patiently for answers, and collaborate with the multilingual learner to address the conception. In this case, it is the language around the teaching and learning about equivalence that presents both challenges and opportunities.

If you want to read more about students' conceptions when learning multiplication, see Lannin, Chval, and Jones (2013). For other mathematical topics, see Appendix A for recommendations. See Appendix B for additional answers for Figure 7.1.

## ANALYZING ADDITIONAL STUDENT WORK

In working with multilingual learners from fifth-grade classrooms, we saw a range of approaches to solving mathematics problems. Before you examine four different examples of fifth-grade multilingual learners' work, take time first to solve the problem students were posed in Try It! 7.2.

**Try It! 7.2**

Solve.

Antonio walked $\frac{3}{10}$ of a mile in the morning. He then walked $\frac{2}{5}$ of a mile after school.

Antonio said he walked $\frac{5}{15}$ of a mile. Is he correct? Explain your thinking.

As you examine student work in Figure 7.3, think about:

❯ How would you describe each student's approach?

❯ How would you respond to each student?

**Figure 7.3**   Student Work From Fraction Problem

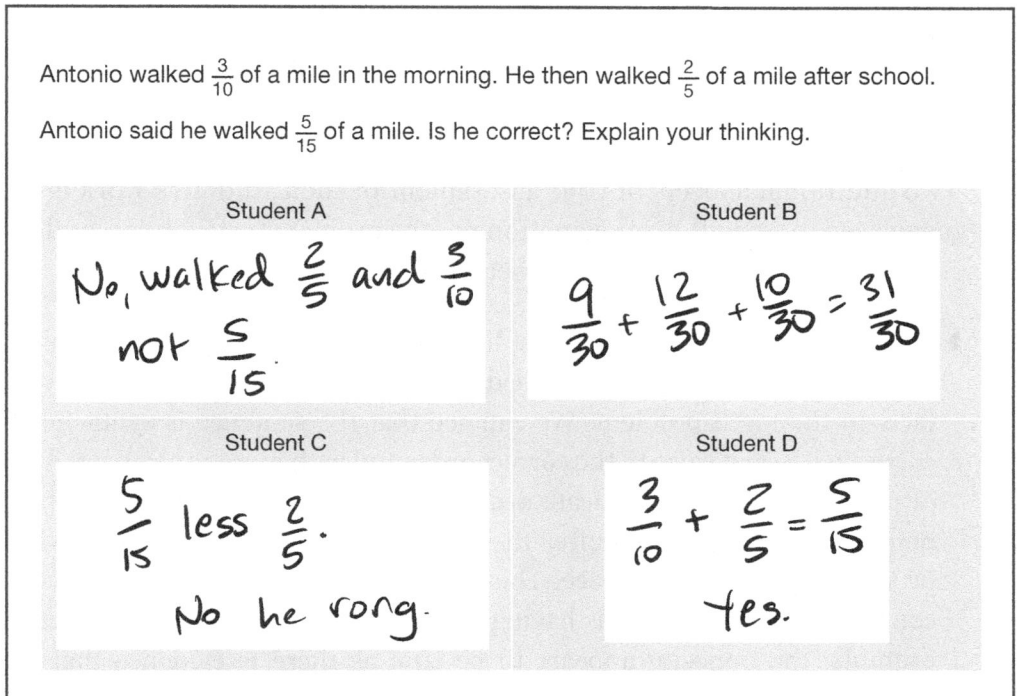

Antonio walked $\frac{3}{10}$ of a mile in the morning. He then walked $\frac{2}{5}$ of a mile after school.

Antonio said he walked $\frac{5}{15}$ of a mile. Is he correct? Explain your thinking.

| Student A | Student B |
| --- | --- |

No, walked $\frac{2}{5}$ and $\frac{3}{10}$ not $\frac{5}{15}$.

$\frac{9}{30} + \frac{12}{30} + \frac{10}{30} = \frac{31}{30}$

| Student C | Student D |
| --- | --- |

$\frac{5}{15}$ less $\frac{2}{5}$.

No he rong.

$\frac{3}{10} + \frac{2}{5} = \frac{5}{15}$

Yes.

Sometimes it is easy to assume students don't understand a problem if they didn't answer the question the way we intended. However, this isn't always the case. As you analyzed each student's work, you may have noticed:

❯ Student A is technically correct and does what the problem asks (i.e., explains thinking), although this may not be an answer commonly expected by teachers. Student A provides a case for insufficient information, meaning it is unclear what the child knows so it would require a discussion with the child to determine a more in-depth assessment.

❯ Student B found a common denominator and writes equivalent fractions for the three fractions in the problem. This work displays knowledge of equivalency, equivalent fractions, common denominators, likely proportional reasoning, and addition of fractions. Although Student B demonstrates a lot of knowledge with unlike denominators, Student B doesn't answer the question that was posed. Thus, it is important to assess this student's understanding further. To do this, you could reflect on questions such as these: Did the student understand what the problem was asking? What question did Student B answer?

▶ Student C is correct. Student C points out that Antonio cannot be correct because one of the addends, $\frac{2}{5}$, is greater than Antonio's proposed solution. Note that the directions do not require determining the sum of $\frac{3}{10}$ and $\frac{2}{5}$.

▶ Student D displays evidence of a common student conception related to the addition of fractions, adding the numerators and denominators respectively to determine the sum (Barnett-Clarke, Fisher, Marks, & Ross, 2010). Student D may have the same conception as Antonio. To investigate that possibility, Student D should be given some additional problems. If you want to read more about students' conceptions about fractions, see Chval et al. (2013).

Two additional aspects of your assessment of each student's work is the language used to provide an explanation per the problem instructions, and the assumptions that you bring as you observe the student's answer.

▶ Student A provides the lengthiest linguistic sample noting that Antonio is wrong and reiterating the specific distances that Antonio walked given the facts in the math problem. We can see that the sentence is syntactically correct with each word in the correct order and with more sophisticated use of punctuation to also indicate order of ideas. Student A omits Antonio's name or the pronoun *he* after the word *no*, which may be an oversight or simply done to save space. This omission is not sufficient evidence to conclude that the student is having language difficulties. In this particular example, the opposite appears to be true as there is evidence that the student has internalized many rules of the English language.

▶ Student B draws on mathematical representation in fractions alone to demonstrate thinking and problem solving. It is important not to assume that this means this child cannot verbally explain ideas in English. This may or may not be the case. Conferencing with Student B or providing more time and space for Student B to write out an explanation, using both the student's dominant language(s) and English, would provide more data with which to make an accurate assessment about the role of mathematical thinking and problem solving for this student.

▶ Student C provides a syntactically correct answer in terms of word order. We can see that the verb has not been included and the word *wrong* has been spelled based on phonetic knowledge (i.e., in U.S. English the word *wrong* is often pronounced "raang" with emphasis on the *r* sound). This child is in a developing stage of second language acquisition and is connecting the sounds of English to their written representations. This can provide important linguistic details as you interact with Student C regarding how this learner came to this mathematical conclusion.

▶ Student D combines reliance on mathematical representation to show work with linguistic pithiness (i.e., the single word *yes*) to indicate that Antonio is correct. However, we cannot assume that because the answer is incorrect.

It is also important to avoid assuming that students' second language competencies are developing at the same rate as their mathematical knowledge. Conferencing or providing space during the class discussion is vital to using mathematics problems such as these to develop language competencies and academic content competencies.

## STRATEGIES FOR DISCERNING BETWEEN MATHEMATICAL AND LANGUAGE ISSUES

Discerning distinctions among mathematical and language issues associated with student work is complex. For instance, as we note in Chapter 3, in many U.S. schools, English is the primary language of instruction; in some classrooms, it is the only language of instruction. Multilingual learners must then negotiate mathematical meanings in an additional language—sometimes translating what others have said into their primary or dominant language, constructing a response, and then translating that response back into English for speakers who do not speak any language other than English. This process can take time. It is important, then, to avoid assuming that a multilingual learner does not know the mathematics or English very well. Patience is key, as are other strategies. In this vein, the following are some strategies you can use when working with multilingual learners, regardless of their first language(s), to facilitate this process and also navigate the practical issues regarding time management in classroom teaching:

▶ Problems can be adapted to assess students' mathematical and linguistic understandings. For example, numbers can be changed to increase or decrease problem difficulty; language can be amplified or made more complex; problem type can be changed; or the problem situation can be retold with names, objects, and contexts with which the multilingual learner is familiar. If students can demonstrate they understand the problem situation, but struggle with the computation, they may benefit from using a calculator or manipulatives (see Chval & Hicks, 2012; Chval & Reys, 2003).

▶ Since many mathematics problems require critical or higher-order thinking, it's important to provide multilingual learners with alternative ways to express their ideas if they face difficulties communicating their ideas. For example, instead of writing a response, students could be asked to share their ideas orally, with the assistance of gestures and drawings, as discussed in Chapter 6.

▶ Their written work only tells part of the story of what students understand. To supplement this, you can watch and listen to multilingual learners as they solve problems to identify how they are approaching problems and probe their thinking (e.g., *Can you tell me how you thought about this problem? Can you tell me/show me how you solved this problem? Can you solve this problem in a different way?*).

▶ Students who demonstrate mathematical competencies but do not answer the mathematics problem may have misinterpreted the task. In such cases, talking with the students can help to identify how they understood the problem.

▶ Understanding how multilingual learners acquire language can support your teaching practice. When analyzing student work, the World-Class Instructional Design and Assessment (WIDA) Consortium resources at https://wida.wisc.edu can help you consider and anticipate what student writing will look like at different developmental levels. Chapters 10 and 11 discuss writing in mathematics classrooms.

## EXAMINING ABIGAIL'S WORK

In addition to examining the work of multilingual learners, teachers must examine the problems they pose and the handouts they provide to students. Imagine you had the handout shown in Figure 7.4.

**Figure 7.4**  "Line Up in Order" Worksheet

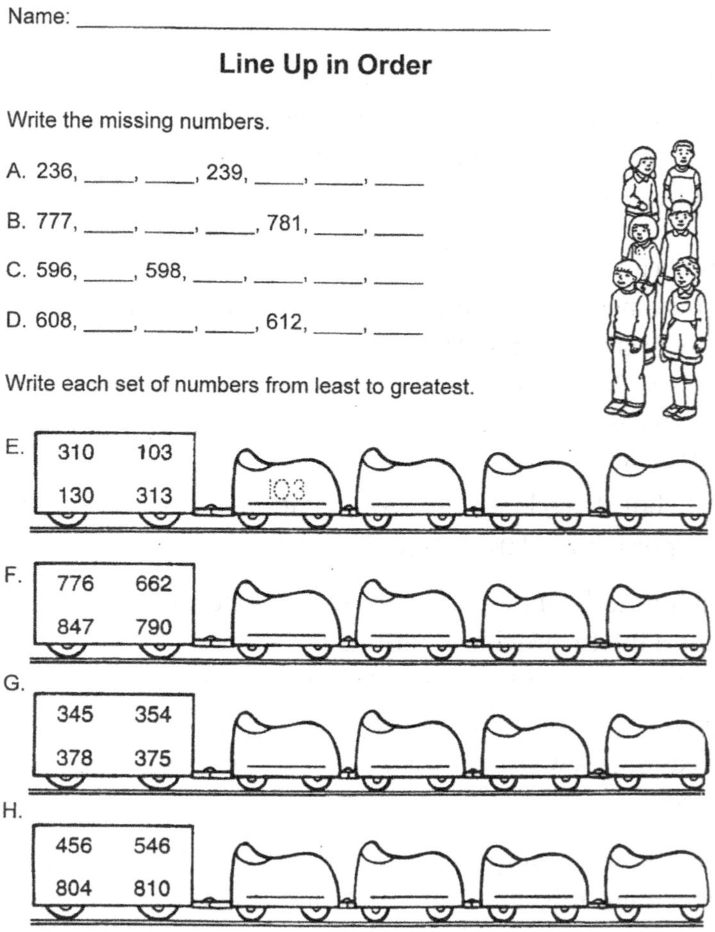

*In addition to examining the work of multilingual learners, teachers must also examine the problems they pose and the handouts they provide to students.*

Name: _____

**Line Up in Order**

Write the missing numbers.

A. 236, _____, _____, 239, _____, _____, _____

B. 777, _____, _____, _____, 781, _____, _____

C. 596, _____, 598, _____, _____, _____, _____

D. 608, _____, _____, _____, 612, _____, _____

Write each set of numbers from least to greatest.

E.  310  103
    130  313

F.  776  662
    847  790

G.  345  354
    378  375

H.  456  546
    804  810

*Source:* Unknown (n.d.).

❱ What challenges could a multilingual learner face when completing this task?

❱ How would you introduce the handout or discuss the content with your students before you distributed it?

In Abigail's third-grade class, her teacher provided students with the handout in Figure 7.4, which was intended to support students' understanding of place value. Examine Abigail's work in Figure 7.5. What understandings does she demonstrate?

**Figure 7.5**   Abigail's Work

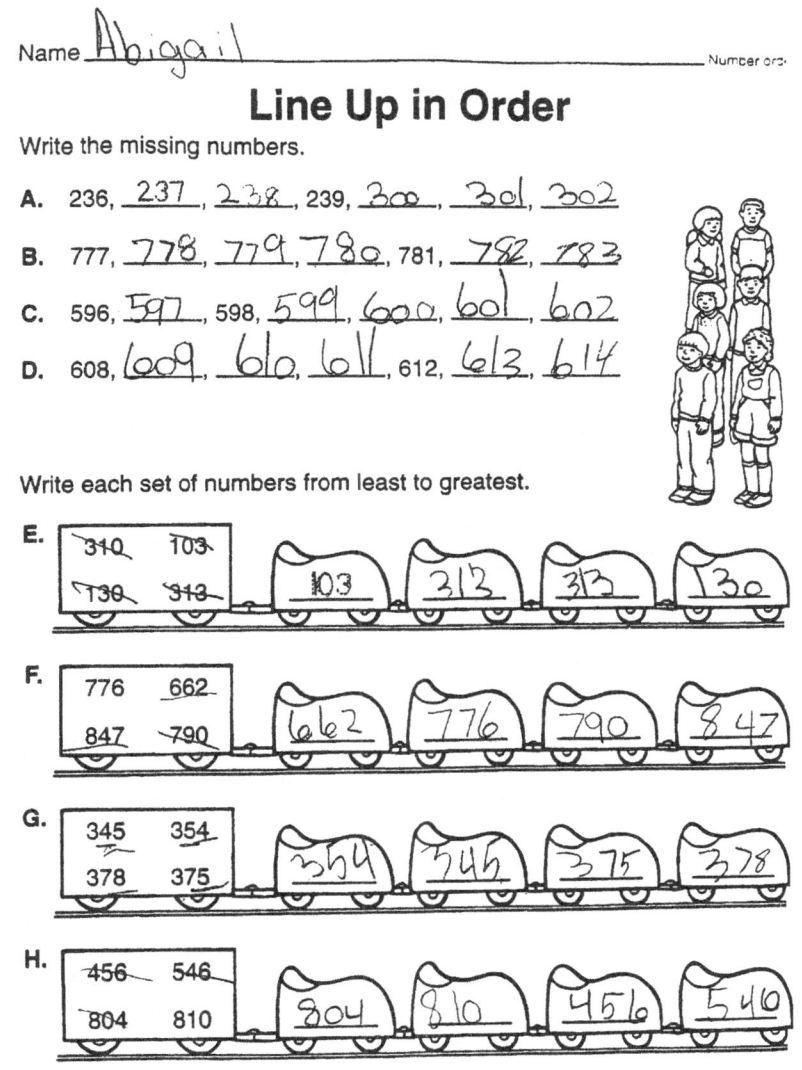

As you know, Abigail's work only tells you part of what she understands. To gain a more robust picture of how Abigail approached problem B, we describe what she did when solving it.

**❝** *Abigail writes 778, then 779. You can hear her teacher in the background saying to the class, "Keep working, guys, you're doing great." She then pauses for 3 seconds. She writes 77 in the third blank and stops. She waits 10 seconds and then begins to look around. She looks back down at her paper and erases the last value she wrote (i.e., 77). She then erases the 8 in 778 and replaces it with a 6 to have 776 as the next number after 777. Next, she erases the 9 in 779, pauses for five seconds, and replaces it with a 5 to have 775 as the number in the next blank. She then writes 774 as the next number in the sequence. (Now on her handout the three blanks after 777 are filled in with 776, 775, and 774.) She spends the next 26 seconds looking at the handout. She then spends 35 seconds looking around the room. Next, she puts her pencil down and begins to look all around her desk (at the floor, underneath the desk, etc.) for 14 seconds.* **❞** *Then, she picks her pencil back up and begins to look at problem E.*

## STOP AND THINK

Stop and think about the mathematical knowledge that Abigail demonstrates.

- What new knowledge do you have about what Abigail understands mathematically after learning about her approach when trying to solve problem B?

- Based on the description of her approach for problem B, what steps would you take next with Abigail?

You may wonder what the teacher was doing while Abigail worked. In this lesson, the teacher visited Abigail to see how Abigail was approaching the worksheet; however, Abigail had erased everything she wrote for problem B. Therefore, her teacher did not recognize that Abigail was struggling to fill in the blanks. Based on what Abigail had written, the teacher decided to bring over base ten blocks. The teacher asked Abigail to model 777 and 778 using the blocks, then showed her how to write a representation of the blocks for those values. Next, the teacher asked Abigail to do the same thing for 779. Abigail did not have any difficulty representing 777, 778, or 779 with the blocks. The teacher encouraged her and walked away. Abigail's teacher did not know Abigail had already written 777, 778, and 779, and then erased them because she was stuck on what followed. Her teacher was also not familiar with research (e.g., Fuson, 1998; Steffe & Cobb, 1988) that found a common place value challenge for students is at decade numbers (i.e., 780 in this case). Abigail's teacher was dedicated, committed, and invested in student success. This was the main reason she joined the research study from which this example is drawn. She used the video data from this interaction with Abigail to analyze both Abigail's experiences and her own misunderstanding of where Abigail floundered in problem solving. Abigail's teacher noted that she had walked away from Abigail before truly understanding the situation, and thus her use of the place value blocks did not actually move Abigail forward mathematically. Abigail's teacher used this insight to ensure that in the future she carefully evaluated what multilingual learners were doing as they encountered new mathematics problems, and provided space for these learners to explain their challenges rather than assuming what the challenges were based on limited evidence.

It was also through interactions such as these that Abigail's teacher began to understand that multilingual learners may need space, time, and careful questions posed by the teacher in order to express any obstacles they face. This approach can also bridge cultural differences such as in cases where a multilingual learner comes from a culture where the teacher is viewed as the preeminent authority in the classroom, or where an authoritative instructional model is used. In such instances, multilingual learners may find it difficult to draw attention to themselves by expressing doubts or struggles, especially in an additional language. Thus, it becomes vital that teachers pose questions to these learners rather than waiting for them to initiate, and also to make space and time for multilingual learners to express their experiences and questions they have as they engage in classroom mathematical learning.

It's also important to recognize that Abigail's understanding of place value was still developing as evidenced by the challenges she faced completing the handout (e.g., moving to the next decade). For Abigail to be successful on such tasks, she must understand that collections of 10 can be treated as a unit (Fuson, 1998; Steffe & Cobb, 1988) and represented symbolically (Hiebert & Wearne, 1992). To support Abigail in developing understanding of place value concepts, it is recommended she use different physical and written representational models alongside multidigit addition and subtraction problems (Carpenter, Fennema, & Franke, 1996; Hiebert & Wearne, 1992).

Fortunately for Abigail's teacher, she learned about this common conception of place value through participating in our intervention and was able to adapt her instruction to provide targeted support to address this conception. For other teachers, a multitude of resources are available that identify common conceptions children have when learning mathematics and research-based strategies to address those. For additional resources, see Appendix A.

## EXAMINING JANESSA'S WORK

Janessa, a fourth-grade multilingual learner, was given the following task:

1. Circle which fraction is larger and explain why using pictures, words, or numbers.

    a) $\frac{7}{8}$ or $\frac{3}{4}$

Now, examine Janessa's work in Figure 7.6. As you analyze it, consider:

▶ How would you describe Janessa's strategy for solving the problem?

▶ When does Janessa's strategy work? Not work?

**Figure 7.6**   Janessa's Work

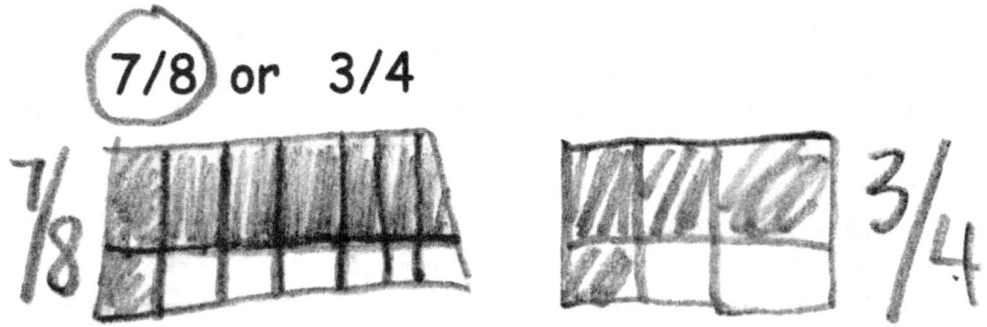

It is important to understand that Janessa uses a strategy that makes sense to her and she uses it consistently. However, she used the same strategy with most other problems that she was given as well. She constructs a two-row array with the number of columns to represent the value of the numerator (e.g., 7). Then, she shades the number that represents the denominator (e.g., 8). Next, she counts the number of unshaded cells to determine the larger fraction (e.g., 6). In this case, 6 is larger than 2 unshaded cells in her second rectangle so she determines the representation for $\frac{7}{8}$ is larger. Janessa's strategy is *very* effective, so she frequently gets the correct answer. However, Janessa's model does not appropriately reflect fractional relationships. For example, she doesn't recognize that comparing fractions using rectangular representations requires that the rectangles (whole unit) are equal in size. The two rectangles Janessa draws are not the same size.

Janessa's work further illustrates that only assessing students' work for accuracy is not enough. If this were the case, her teacher would have noticed her mathematical conception (not language challenge) and began an intervention to address it. As her teacher determined what intervention to use with Janessa, her teacher also considered what models Janessa had experienced. It is important to recognize that the mathematical models that students use are strongly tied to curricula and the models their teachers use. The overuse of the area model, and specifically rectangles with two rows, influenced the development of Janessa's conception of representing fractions. It may be useful to ask yourself, "What models do my students see and use?" Most importantly, conceptions like Janessa's must be directly addressed so that students can be successful in future mathematics coursework.

## THE IMPORTANCE OF INTERVIEWING MULTILINGUAL LEARNERS

Interviewing multilingual learners can provide greater depth and detail of their mathematical thinking that cannot be garnered from their written work alone (Castellón, Burr, & Kitchen, 2011). Let's take a look at an example of this from Eduardo, a third-grade multilingual learner. You will find a sample of Eduardo's work in Figure 7.7. Eduardo was given a pre- and posttest in both English and Spanish (his first language), administered approximately two weeks apart (first English, then Spanish). The tests were given one-on-one by an administrator who read words or questions if the student requested and probed to clarify the student's thinking for each test item. As you review Eduardo's work, focus on noticing differences between the English and Spanish versions along with differences between the pre- and posttests.

**Figure 7.7**  Eduardo's Tests

Note: Between the administration of the two posttests, Eduardo's teacher was shifting from the bottom representation on the English posttest (of drawing individual items) to the top representation on the Spanish posttest (of drawing numerals to represent the number of individual items).

| PRETEST | POSTTEST |
|---|---|
| Jackie hizo 24 galletas. Jackie puso 4 galletas en cada plato. ¿Cuántos platos usó Jackie para colocar todas las galletas? | Jackie hizo 24 galletas. Jackie puso 4 galletas en cada plato. ¿Cuántos platos usó Jackie para colocar todas las galletas? |
| $29$ *pluto* |  *Each circle in this drawing represents a plate. The 4 on each plate represents the number of cookies on the plate.* |
| Anna baked 28 cookies. Anna put 4 cookies on each plate. How many plates did Anna use to put all of the cookies on plates? | Anna baked 28 cookies. Anna put 4 cookies on each plate. How many plates did Anna use to put all of the cookies on plates? |
| $32$ *plates* |  *Each circle in top row of this drawing represents a plate. Each small circle represents a cookie. The small circles are grouped by fours, and the arrows indicate which plate they are on.* |

In your examination of Eduardo's work, you may have noticed the changes in his mathematical thinking between the pretests and posttests. For example, on the posttest, Eduardo models the plates from the story problem, which is not found on his pretest. Additionally, on his posttest, his Spanish representation is more abstract than his representation on the English version. You may also have noticed that Eduardo answered 7 for his final answer on his posttest in English, but not in Spanish. Although you may have assumed from his written work alone that Eduardo did not know what the final answer was based on his Spanish posttest, when asked to describe his problem-solving strategy, he thoroughly explained his mathematical thinking, representation, and understanding of the final answer of 6 plates to the administrator. Based on his work alone, it is impossible to know this. If you were Eduardo's teacher, you might consider next asking Eduardo to solve a similar problem as the Spanish posttest and watch him solve it to understand his strategy. If Eduardo was unable to explain his thinking in English, a colleague who was fluent in Spanish could probe his thinking. For multilingual learners, sometimes teachers may intuitively believe a student will be more successful on a problem if it's translated; however, this isn't always the case.

*Sometimes teachers may intuitively believe a student will be more successful on a problem if it's translated; however, this isn't always the case.*

## THINKING ABOUT ANALYZING MULTILINGUAL LEARNERS' MATHEMATICAL WORK IN YOUR PRACTICE

In this chapter, you analyzed the mathematical work of multilingual learners to identify and refine your ability to discern distinctions among mathematical and language challenges. In addition, you explored how

- watching multilingual learners solve problems in real time can help you assess their mathematical understandings;

- analyzing multilingual learners' mathematical work is critical to facilitating their mathematical development;

- examining written work to identify patterns, conceptions, and understandings can provide insights for your instruction;

- examining multilingual learners' mathematical work only tells a part of the story of their thinking—sometimes it's not enough to only look at the student's final work before determining the next instructional steps; and

- encouraging multilingual learners to share their thinking and approaches to mathematics problems helps you assess what they know.

As you analyze and reflect on multilingual learners' mathematical work, it may be beneficial to ask yourself questions, such as the following:

▶ Are there problems with the language or context of the problem? Is this a high-quality problem or task?

▶ Did I present this concept in a way that students can access? Did I effectively use visuals or contexts?

▶ How engaged are the students? Am I holding them accountable to the same expectations as peers, or am I allowing them to sit on the sidelines?

▶ Are the students being effectively challenged?

▶ Are the students using appropriate strategies to solve the mathematics problem?

▶ In what ways are the students communicating their thinking? What language resources are the students using to communicate mathematically?

▶ Can the students appropriately represent the problem or task?

▶ Can the students solve the problem if it's read to them?

▶ Can the students solve the problem in their preferred language?

▶ Do the students understand what the problem or question is asking?

▶ Do the students understand the mathematical concept required to be successful?

▶ What can I ask individual students to better understand their thinking?

▶ How can I extend students' thinking? How can I confront unproductive conceptions?

## Reflect

- What is important for you to consider as you analyze the mathematical work of multilingual learners?

# CHAPTER 8
# INVESTIGATE MEANINGS TO ENHANCE MULTILINGUAL LEARNERS' LANGUAGE DEVELOPMENT

## Key Concepts

In this chapter, you will

✓ understand the importance of developing specialized mathematical language with multilingual learners.

✓ examine multiple meanings associated with language.

✓ identify instructional strategies to distinguish between everyday language and specialized mathematical language.

✓ identify instructional strategies to transition students' use of everyday language to specialized mathematical language.

✓ understand the importance of amplifying academic language to support students as they build meaning.

Over the course of our careers, there have been multiple instances where we have introduced or used a word or phrase during our mathematics instruction only to learn later in the lesson that students were thinking of entirely different meanings. For example, during one lesson, a student wrote "kind of pets," referring to a graph of different types of pets. His multilingual partner said that "*kind* means 'nice.'" This led us to think more about the ways we craft our instruction to facilitate multilingual learners' transition from familiar to specialized mathematical language. To aid you in thinking more about this, you will explore two forms of specialized mathematical language in this chapter: (1) language that has both mathematical and everyday meanings, such as *similar* or *congruent*, and (2) language that primarily has mathematical meanings, such as *algebra*, *hypotenuse*, or *arithmetic*.

## THE IMPORTANCE OF ACADEMIC LANGUAGE FOR MULTILINGUAL LEARNERS

Imagine you are a third-grade multilingual learner who attends a classroom in a small town where the majority of students are English speakers. You can establish basic communication, but you are still learning the meanings of the language and adapting to the culture. In your mathematics class, the teacher is talking about rounding up. Although you may know the meaning of *round* and *up*, this is the first time you have been exposed to the expression *rounding up*. What different meanings could you associate with this expression?

Rounding up is specialized mathematical language. Specialized mathematical language is the specific language used to refer to mathematical concepts and processes. In contrast, everyday language is used in daily interactions.

### STOP AND THINK

Stop and think about what strategies you have used (or would use) to explain specialized mathematical terms, such as *rounding up*.

- Now, imagine a colleague said to you that specialized mathematical language was too complex to be taught to multilingual learners. How would you respond?

It's important for mathematics teachers to recognize that language in mathematics is complex. The solution to this complexity is not to simply avoid mathematical academic terms and phrases, as avoiding this complexity with multilingual learners restricts their access to learning academic language and academic content that they need to engage actively in the classroom.

## WHAT THE RESEARCH SAYS ABOUT ACADEMIC LANGUAGE

Since 1991, the National Council of Teachers of Mathematics has advocated for the value and importance of communication for mathematics learning. Moreover, research shows that every student benefits from explicit inclusion of academic language in instruction and curriculum (Genesee, 2006; Gibbons, 2015). Academic language is not limited to vocabulary; it also comprises a set of intellectual and social practices shaped by cultural norms and values, such as understanding different meanings and using expressions in context, comprehending how language is used in academic discourse, and understanding complex sentence structures and syntax along with the cognitive demands of their meanings (Gottlieb & Ernst-Slavit, 2019; Valdés, 2004; Zwiers & Soto, 2017). In the classroom, teachers must attend to the ways they facilitate multilingual learners' acquisition of academic language to ensure they can be mathematically successful (MacDonald, Lord, & Miller, 2019).

Unfortunately, we do not learn language through "osmosis" or by solely interacting with fluent or native-like speakers; teachers must explicitly teach it (de Jong & Harper, 2005; de Jong, Harper, & Coady, 2013). However, this does not mean that teachers should focus solely on teaching particular words or phrases in isolation, as this is unproductive for student learning and an inefficient use of instructional time (Gibbons, 2015). Said another way, mastering words or phrases alone does not ensure students can effectively engage in mathematical discourse (Moschkovich, 2002). As a result, teachers must attend to the ways they enhance multilingual learners' acquisition of specialized mathematical language and provide opportunities for language to be used in and across contexts. This instruction must also include opportunities for students to negotiate meaning while drawing on their prior experiences and knowledge (de Jong & Harper, 2005; Zwiers & Hamerla, 2018).

*Teachers must attend to the ways they enhance multilingual learners' acquisition of specialized mathematical language and provide opportunities for language to be used in and across contexts.*

## DISTINGUISHING BETWEEN MULTIPLE MEANINGS OF LANGUAGE

It is critical for teachers to ensure language is not a barrier for students' learning of mathematical concepts. As teachers facilitate students' use of academic language, there are different challenges teachers need to take into account, such as the following:

- Some language has multiple meanings, but not mathematical meanings (e.g., *leaves the room* vs. *leaves on a plant*).

- Some language has multiple meanings, including meanings used in mathematics (e.g., *change in your pocket* vs. *rate of change*).

- Some language is specific to mathematics (e.g., *Pythagorean theorem, parallelogram*).

As teachers plan lessons, they must consider what language students may be familiar with that can be drawn on during instruction and be used as a tool to distinguish between everyday and mathematical meanings.

## STOP AND THINK

Stop and think about how often you ask your students if they have heard of a specific word or phrase or know what it means.

• Now imagine you were going to introduce a third-grade lesson where understanding *roll* in the context of bundling T-shirts was critical, as we saw in Context 5.3 in Chapter 5. What are different meanings for the word *roll* that third graders could identify?

Let's peek into the classroom of Ms. Bristow—a third-grade teacher whose classroom we've visited in prior chapters—and see how she acts as a mediator to bridge students' everyday and academic language. In this thematic unit, Ms. Bristow began by reading the children's story "Grandma Eudora's T-Shirt Factory" by Cathy Twomey Fosnot (2007). In the story, a family creates bundles, or "rolls," of T-shirts in amounts of 10, which are later connected to base ten blocks. As you read Transcript 8.1, pay attention to how Ms. Bristow distinguishes between everyday uses of *roll* and the particular academic use of *roll*. It's also important to know that *roll* in English and *rollo* in Spanish are cognates. Transcript 8.1 begins immediately after Ms. Bristow finishes reading the story. On the board, she projects Figure 8.1 along with the word *roll*.

**Figure 8.1**   Images Ms. Bristow Projected on the Board

*Source:* Dinner roll cislander/iStock.com; Coin roll TCassidy/iStock.com; Roll downhill Marc Dufresne/iStock.com; T-shirt roll BobChaiyuth/iStock.com

**Transcript 8.1**

| Ms. Bristow: | I've heard the word *roll* a lot. I don't always think about a roll of T-shirts, though, when I think about rolls. I think about different things. I think about dinner rolls. Have you ever had a roll at Thanksgiving? [*Points to a picture of a roll on the SMART Board®*] |
| --- | --- |
| Students: | Yeah. |
| Ms. Bristow: | What is a roll, Lyra? |
| Lyra: | It's like bread. |
| Ms. Bristow: | Yeah, it's a little piece of bread, right? But, we are not thinking about dinner rolls, are we? |
| Students: | No. |
| Ms. Bristow: | No, I didn't say anything about dinner rolls [in the story]. What about this? What kind of rolls are these? [*Points to a picture of rolls of coins*] Brian, what do you think those are? |
| Brian: | Rolls of coins. |
| Ms. Bristow: | Rolls of coins, and we've talked about rolls of coins before. Are they rolling up coins in this book? |
| Students: | No. |
| Ms. Bristow: | No, not rolling coins. What about—you can roll down a hill, right? |
| Students: | Yeah. |
| Ms. Bristow: | Is that the kind of roll they are talking about? |
| Students: | No. |
| Ms. Bristow: | No. What kind of roll are they talking about? |
| Bruno: | Rolling T-shirts. |
| Ms. Bristow: | Bruno, what is it? |
| Bruno: | Folding T-shirts. |
| Ms. Bristow: | Yeah, so they are folding T-shirts, and then they are making a roll. So, here is somebody—they must be packing for a trip. They got a set of T-shirts, and they take them and kind of roll them into a little bundle, and that's a roll. That's what they're talking about. How many T-shirts does Nicholas the organizer [from the story] put in one roll? How many? Show me with your hands. How many are there? Are there 5? I'm seeing some 5s. Are there 10? Are there more? Portia, how many are there in the roll? |
| Portia: | 10. |
| Ms. Bristow: | So, there are 10 T-shirts in a roll. Now, so I think we know that there are 10 T-shirts in a roll, and I've got some shirts that we can roll, too. So, we can roll some shirts later. |

*Source:* Chval, K. B., Pinnow, R. J., & Thomas, A. (2015). *Mathematics Education Research Journal*. Used with permission.

# STOP AND THINK

Stop and think about Transcript 8.1.

- What are some strategies that Ms. Bristow uses to distinguish between everyday uses of *roll* and the particular academic use of *roll* in the story?

- What are some ways you can distinguish between specific academic meanings of language, particularly words that may hinder mathematical success?

# INTRODUCING SPECIALIZED MATHEMATICS LANGUAGE

Teachers use their language to simultaneously model the language of a mathematician while acting as a mentor for students. However, it's not enough to simply use academic language in front of students; teachers must facilitate students' acquisition of academic language. Let's peek into the classrooms of Ms. Martínez and Ms. Bristow to see how they introduced the specialized mathematics language of *quadrilateral*, *remainder*, and *congruent* and facilitated their students' acquisition of this language.

## Ms. Martínez: Quadrilaterals

Ms. Martínez recognizes that her students may be unfamiliar with specialized mathematical language. As a result, she populates her lessons with academic language on the first day of school. (Remember that at the time of our classroom observation Ms. Martínez had 20 years of teaching experience and the student population in her urban school was 100% multilingual learners, but the practices she implemented can be used by teachers with a range of experience and with multilingual learners from other linguistic and cultural backgrounds.) Not surprisingly, Ms. Martínez's language differed at the beginning of the year (as shown in Transcript 8.2) from the end of the year when students used mathematical language more fluently (Razfar, Khisty, & Chval, 2011).

In Transcript 8.2 (from Khisty & Chval, 2002), Ms. Martínez is in the midst of teaching a lesson on geometry—a topic she used at the beginning of every school year due to its visual nature—during the second mathematics lesson of the school year. As you read, pay attention to the way she introduces the term *quadrilateral* and other steps she takes to foster her students' language development.

**Transcript 8.2**

| | |
|---|---|
| Ms. Martínez: | *It has four sides.* You know what? I'm going to put the word *rectangle* into a *category*. And I am going to call this category *quadrilaterals*. [*Writes "quadrilaterals" on the board*] Do you recognize or at least listen to the sound of the word and see if there is any part of this word that you recognize? *Qua–dri–lat–er–al. Qua–dri–lat–er–al. Qua–dri–lat–er–al. Cuadro*, right? What is a cuadro? |
| Sofia: | A square. |
| Juan: | A shape. |
| Ms. Martínez: | A shape that has . . . ? |
| Lina: | Four sides. |
| Ms. Martínez: | A shape that has four sides. Look at your classroom. *Do you see a lot of shapes that are quadrilaterals?* |

*Source:* Khisty, L. L., & Chval, K. B. (2002). *Mathematics Education Research Journal.* Used with permission.

## STOP AND THINK

Stop and think about Transcript 8.2.

• What do you notice about the way Ms. Martínez introduces the term *quadrilateral*?

• What are some actions or steps Ms. Martínez uses to foster students' language development?

Khisty and Chval (2002) wrote:

> [It] is not so simple as to simply give words to students. The words represent meanings that are waiting to be developed and eventually internalized. Therefore, which words are presented to the students and how they are developed are vitally important. Just as important is that students have opportunities to use these words in their talk and as they work. (p. 155)

Examine the transcript one more time in Figure 8.2. In the right-hand column, highlight aspects of Ms. Martínez's language that you did not notice when you read it the first time.

**Figure 8.2**   Ms. Martínez's Actions in Relation to Transcript 8.2

**T: Ms. Martínez**

| TRANSCRIPT | MS. MARTÍNEZ'S ACTIONS |
|---|---|
| 1.  T: *It has four sides*. You know what? | 1.  Restates a key attribute of the shape being discussed (i.e., rectangle). |
| 2.  I'm going to put the word *rectangle* | 2.  Restates the name of the object being discussed, *rectangle*. |
| 3.  into a *category*. | 3.  Introduces the concept and formal term *category*. |
| 4.  And I am going to call this category | 4.  Restates *category*. |
| 5.  *quadrilaterals*. [*Writes "quadrilaterals" on the board*] | 5.  Introduces the name of the category, *quadrilaterals*, and documents it on the board. |
| 6.  Do you recognize | 6.  Uses a term with a Spanish cognate (i.e., *recognize/reconocer*) and uses sophisticated language. |
| 7.  or at least listen to the sound of the word | 7.  Asks students to listen to the sound of the word. |
| 8.  and see if there is any part of this word that you recognize? | 8.  Asks students if they are familiar with part of the word. |
| 9.  *Qua–dri–lat–er–al. Qua–dri–lat–er–al. Qua–dri–lat–er–al.* | 9.  Restates the term by sounding out syllables *three* times. |
| 10.  *Cuadro*, right? | 10.  States the cognate from Spanish. |
| 11.  What is a cuadro? | 11.  Asks students what *cuadro* means, referring to *cuadrado* in Spanish or *square* in English. |
| 12.  Sofia: A square. | 12.  Ignores Sofia's response. |
| 13.  Juan: A shape. | 13.  Makes a decision to take up Juan's response. |
| 14.  T: A shape that has . . . ? | 14.  Revoices Juan's response and elicits the key attribute from line 1. |
| 15.  Lina: Four sides. | 15. |
| 16.  T: A shape that has four sides. | 16.  Combines Juan's and Lina's responses. |
| 17.  Look at your classroom. *Do you see a lot of shapes that are quadrilaterals?* | 17.  Elicits students to notice *quadrilaterals* in their classroom to make visual connections to the new term. |

You may have noticed that Ms. Martínez scaffolds the specialized mathematical language for her students. For instance, she introduces the word and concept *category*, introduces the mathematical word *quadrilateral*, and then puts the word in context in a way that invites students to use it. It also is interesting how she implicitly capitalizes on students' knowledge of Spanish (by connecting it to *cuadro*) to have them construct a meaning for *quadrilateral* (Khisty & Chval, 2002). For teachers whose students speak a first language other than Spanish, a quick internet search can provide first-language terms that the teacher can draw on during instruction. Multilingual children's family members can also be a source of first-language words and phrases that are very helpful for enhancing instructional language and bridging cultures in the classroom.

Ms. Martínez uses a rhetorical device in her opening discussion with students (i.e., "You know what?"), which we know multilingual learners notice and use themselves in talk with their peers (Pinnow & Chval, 2015). From a positioning perspective, this is important because it demonstrates the confidence Ms. Martínez has in her students' ability to learn specialized mathematical language while also positioning them as competent learners.

## Ms. Bristow: Remainder

Teachers can act as mediators between everyday and academic language to leverage students' daily language to introduce specialized mathematical language. Let's look at how Ms. Bristow does this in Transcript 8.3 (from Pinnow & Chval, 2014) when she introduces the mathematical term *remainder*.

**Transcript 8.3**

| | |
|---|---|
| Ms. Bristow: | Well, you know, sometimes, there are things that are left over or extras, right? So we had this one extra. And sometimes when you are dividing things, there are going to be times when you have something left over. [*The SMART Board® displays "Times when you have leftovers" and "Leftover—something you have that you haven't used or finished"*] |
| Ms. Bristow: | So sometimes we have an extra, right? I heard Lance say the word *extra* [*writes the word* extra *next to the word* leftover *on the SMART Board®*]. So what are the times you have something extra or left over? So tell me about a time, just in your day, where you had something extra or left over. Mary, what do you think? [*As each child shares an idea, Ms. Bristow writes the child's name and idea on the SMART Board®*] |
| Mary: | One time in dinner, there was a leftover hotdog. |
| Ms. Bristow: | Okay, so Mary says a hotdog from dinner. So, we think of a leftover being something extra. So a leftover is like something you had that you didn't use or didn't finish. Wayne, is there a time you have had a leftover? |
| Wayne: | Last night we had leftover steak for dinner. |
| Ms. Bristow: | Okay, so you ate something that had been left over, right? |
| Marty: | The play we did from Christmas—there weren't a lot of people, so we had to add extra characters, so you gave some characters to Casey. |
| Ms. Bristow: | Extra scripts? |

(continued)

(continued)

| | |
|---|---|
| Students: | Yes. |
| Ms. Bristow: | So scripts from the Arthur play. Okay, Maria? When have you had a leftover? |
| Maria: | When we are thinking we are going to spend an amount of money, but we have leftover money. |
| Ms. Bristow: | So, if you go to the grocery store and you have $5 to spend, and maybe you spent $4.95 and you have one nickel left or 5 cents left. So you might have money left over from buying something. People use different words for leftovers. Sometimes they say, "We have this extra," like, at lunchtime, do you guys ever have a leftover? I have seen some crispy rice cereal treats; those are leftovers from lunch, aren't they? So there are extras you didn't eat. Or sometimes people will move their snacks into the middle of the table at lunch as extras. Sometimes people use this word, *leftover*, and sometimes people use this word, *extra*, and sometimes people use this word, *remainder*. [*Ms. Bristow writes the word* remainder *next to the words* leftover *and* extra *on the SMART Board®*] |
| Craig: | Remain. |
| Ms. Bristow: | Oh, and so there is this word, *remain*, which is what is still left. Okay? |
| Craig: | Like, what is remaining of it. |
| Ms. Bristow: | Yes, like what is remaining of it on the table. So sometimes people use this word *remainder*, and that is actually the mathematical word that people use. So instead of saying *extra*, which really means the same thing, *remainder* is, kind of, the way we usually talk about it in math, but it means the same thing as leftovers or extras. |

## STOP AND THINK

Stop and think about Transcript 8.3.

- What are some strategies that Ms. Bristow uses to leverage students' everyday language and prior experiences to introduce *remainder*, a specialized mathematical term?

## Ms. Martínez: Congruent

Imagine you were going to teach a fifth-grade lesson centered on the task in Figure 8.3 and were going to introduce the term *congruent*. Unlike *remainder, congruent* is a term not often used in everyday language, and students are likely to have limited prior knowledge of it when they first experience it in a mathematics classroom.

**Figure 8.3**  Task About the Area of a Right Triangle

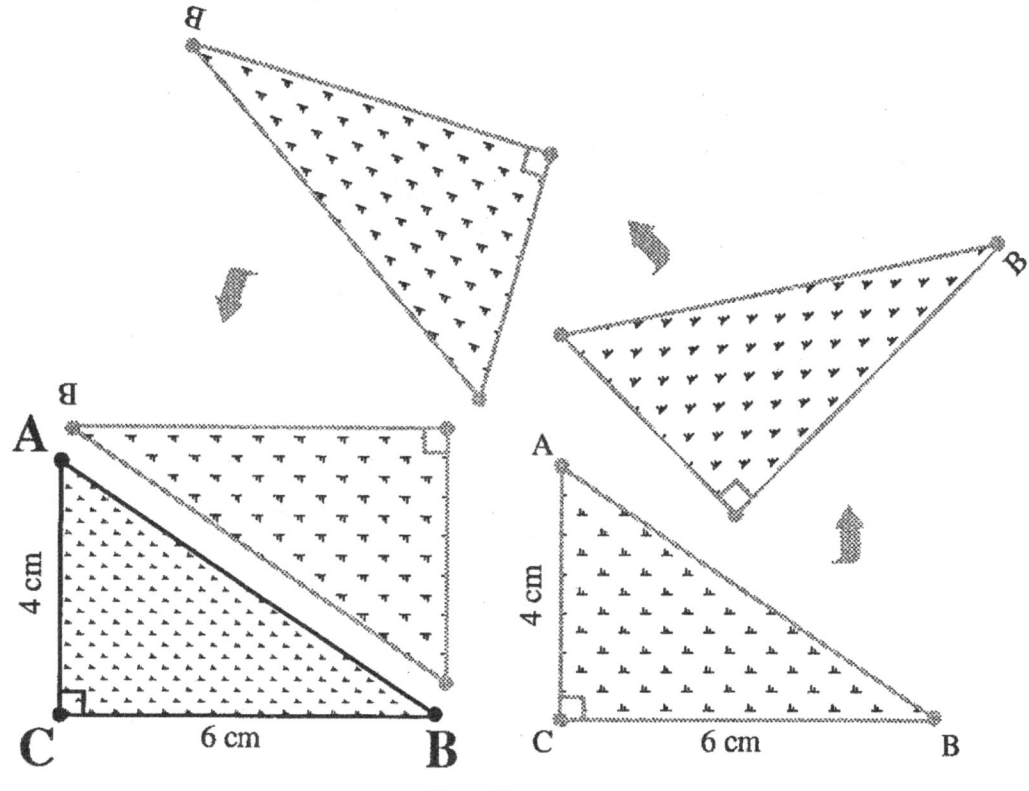

*Area of Right Triangles*                                                                 *3*

$\triangle \mathbf{ABC}$ plus $\triangle$ABC has twice the area of $\triangle \mathbf{ABC}$. $\triangle$ABC is turned up-side-down to form a rectangle with $\triangle \mathbf{ABC}$.

Figure F

3. The triangle and its congruent copy make a rectangle 6 cm by 4 cm. The rectangle is shown below. Find its area.

   Area of the rectangle _____ sq. cm

4. Now find the area of $\triangle \mathbf{ABC}$.
   Hint: We only want half of the rectangle's area.

   Area of $\triangle \mathbf{ABC}$ _____ sq. cm

Figure G

© *David A. Page, Philip Wagreich*                                    *Maneuvers with Triangles*

*Source:* Page, D. A., Wagreich, P., & Chval, K. (1993). *Maneuvers with triangles*. Palo Alto, CA: Dale Seymour. Reprinted with permission.

## STOP AND THINK

Stop and think about the task in Figure 8.3.

- What challenges may the expression *congruent* present for multilingual learners that differ from those presented by language such as *remainder*?

- How would you introduce *congruent* to your students?

Ms. Martínez's fifth-grade class completed the task in Figure 8.3. Now, let's take a look at how Ms. Martínez introduced *congruent* to her fifth graders. As you read Transcript 8.4 (from Chval & Khisty, 2009, pp. 138–139), think about how Ms. Martínez connects everyday language to the mathematical concept of congruence.

**Transcript 8.4**

| | |
|---|---|
| Ms. Martínez: | What do I need to do to the 24, to get the area of that right triangle? |
| Juan: | Divide by 2. |
| Ms. Martínez: | Why do I divide it by 2? |
| Mateo: | You have two triangles. |
| Ms. Martínez: | I have two *congruent* triangles here. Two *equal* parts, two *exact* triangles. I want only the area of my original triangle, ABC. Then I'm going to divide this by 2. And what will my answer be? . . . Number three. Would you please read that, Julia? |
| Julia: | The triangle and its . . . |
| Ms. Martínez: | Congruent. |
| Julia: | Congruent [*struggling*] . . . |
| Ms. Martínez: | Look at that word, everyone. *Congruent.* What does that mean? |
| José: | Like another copy. |
| Ms. Martínez: | An *exact copy.* Because here, look, here is the circle. Is this circle congruent to that circle? |
| Chorus: | No. |
| Ms. Martínez: | No, they're not *exact copies.* They're similar, they're both circles, but they're not *exact copies.* |
| Chorus: | Yes. |
| Ms. Martínez: | How about this one and this one? |
| Chorus: | Yes. |
| Ms. Martínez: | They appear to be congruent to each other. I agree. They appear to be congruent. But this one and this one are not congruent, are they? |
| Chorus: | No. |
| Ms. Martínez: | So, *congruent* means an *exact copy.* José, you are super right. |

*Source:* Chval, K. B., & Khisty, L. (2009). Latino students, writing, and mathematics: A case study of successful teaching and learning. In R. Barwell (Ed.), *Multilingualism in mathematics classrooms: Global perspectives.* Bristol, UK: Multilingual Matters. Used with permission.

As you read Transcript 8.4, you may have noticed Ms. Martínez do each of the following:

- Recognized that students may not have had exposure to the term *congruent* prior to their mathematics class. Knowing this, Ms. Martínez intentionally connected the term *congruent* with students' everyday language of *equal* and *exact* to bridge their understanding.

- Incorporated cognates into her instruction. In Spanish, the cognate *congruente* has a daily-life meaning (i.e., "coherent," "reasonable," or "convenient"), which her Latinx students may have been familiar with. In English, the word *congruent* also means "consistent" (e.g., a congruent version) or "corresponding" (e.g., this person's statements are not congruent with others'); however, its use in English is limited compared to the Spanish cognate.

- Signaled how to reformulate everyday language (i.e., copy) using the academic expression *congruent*. Then, Ms. Martínez used the new language in context and encouraged students to use it, too.

- Revoiced (i.e., restated and clarified) student contributions to move student language acquisition forward. For example, after a student said, "You have two triangles," Ms. Martínez revoiced the student's response to include the term *congruent* (i.e., "I have two *congruent* triangles") to increase its mathematical precision. In this same turn, she used the terms *equal* and *exact* (i.e., "Two *equal* parts, two *exact* triangles") to further define *congruent* for her students.

- Used students' initial conceptions of *congruent* as a bridge to its mathematical meaning. Specifically, Ms. Martínez solicited students for a definition of *congruent* by asking, "*Congruent*. What does that mean?" In response, José provided a new description, "another copy." Ms. Martínez then refined José's response by combining her description with José's to create "exact copy" over multiple turns (i.e., "An *exact copy*," "*exact copies*," and "*congruent* means an *exact copy*"), and she gives credit to José in the last line of the transcript.

To emphasize some of these points and the nuances of Ms. Martínez's practice, we highlight her important actions in Transcript 8.5.

**Transcript 8.5**

| | | |
|---|---|---|
| Ms. Martínez: | What do I need to do to the 24, to get the area of that right triangle? | Ms. Martínez revoiced the student contribution, but included *congruent* to increase its mathematical precision. |
| Juan: | Divide by 2. | |
| Ms. Martínez: | Why do I divide it by 2? | |
| Mateo: | You have two triangles. | |
| Ms. Martínez: | I have two *congruent* triangles here. Two *equal* parts, two *exact* triangles. I want only the area of my original triangle, ABC. Then I'm going to divide this by 2. And what will my answer be? . . . Number three. Would you please read that, Julia? | She also uses the terms *equal* and *exact*—terms her students are familiar with—to further define *congruent* for her students. |
| Julia: | The triangle and its . . . | |
| Ms. Martínez: | Congruent. | Ms. Martínez asks for a definition of *congruent*. |
| Julia: | Congruent [*struggling*] . . . | |
| Ms. Martínez: | Look at that word, everyone. *Congruent*. What does that mean? | |
| José: | Like another copy. | Here José provides a new description, "another copy." |
| Ms. Martínez: | An *exact copy*. Because here, look, here is the circle. Is this circle congruent to that circle? | |
| Chorus: | No. | |
| Ms. Martínez: | No, they're not *exact copies*. They're similar, they're both circles, but they're not *exact copies*. | Ms. Martínez refines José's description to include *exact*. |
| Chorus: | Yes. | |
| Ms. Martínez: | How about this one and this one? | |
| Chorus: | Yes. | |
| Ms. Martínez: | They appear to be congruent to each other. I agree. They appear to be congruent. But this one and this one are not congruent, are they? | Ms. Martínez repeats the refined description. |
| Chorus: | No. | |
| Ms. Martínez: | So, congruent means an *exact copy*. José, you are super right. | Ms. Martínez repeats the refined definition of exact copy for *congruent* again and then gives credit to José. |

*Source:* Chval, K. B., & Khisty, L. (2009). Latino students, writing, and mathematics: A case study of successful teaching and learning. In R. Barwell (Ed.), *Multilingualism in mathematics classrooms: Global perspectives*. Bristol, UK: Multilingual Matters. Used with permission.

From this transcript forward, Ms. Martínez consistently referred to *congruent* as a copy. Then, over the course of the next several lessons, Ms. Martínez combined these words together, *congruent copy*.

> 66 *As students became comfortable using the term, she removed the word "copy" and began to use more precise language, such as "congruent triangle." Eventually, congruent became a word that appeared in the writing and speaking of every student in the classroom. (Chval & Chávez, 2011, p. 263)* 99

Overall, Ms. Martínez guided her students' appropriation of *congruent* by using it frequently in the context of solving problems and by creating meaning for it, while simultaneously using her students' everyday language as resources, not as an obstacle to learning (Moschkovich, 2012). Her practice benefited students' learning and language acquisition more so than if she had simply provided a formal definition or continued to use simplified, less precise language in her instruction.

# AMPLIFIED USE OF ACADEMIC LANGUAGE

Another way to facilitate students' language acquisition is to amplify academic language and encourage students to use it (Gibbons, 2015). For instance, teachers can use whole-class discussions as a way for students to learn *through* language and *about* language since they are collectively constructing knowledge, not just providing answers (Chval, Pinnow, & Thomas, 2015). In discussions about Ms. Martínez's language at the beginning of the school year, some teachers often recognize that most of the classroom dialogue is populated by Ms. Martínez while her students' responses are brief. In analyzing transcripts from the second semester, it should be noted that her students' responses are lengthy paragraphs. However, at the beginning of the year, her students have not had the opportunity to learn the words in English that Ms. Martínez is introducing. Therefore, we do not see evidence of the students' use of these words in English until Ms. Martínez helps them build meaning for those words. Thus, teachers need to consider when to do a lot of talking (e.g., early in the year, introducing new topics) and when to scale back to allow students to take the lead.

Figure 8.4 illustrates the frequency of selected words that Ms. Martínez used in the first 12 mathematics lessons during whole-class instruction (Chval & Khisty, 2009). (Note: This figure does not include instances when she used these words during her interactions with individual students or with small groups.)

**Figure 8.4** Frequency of Ms. Martínez's Words Over the Course of the First 12 Mathematics Lessons

| WORD | TOTAL NUMBER OF TIMES USED | AVERAGE TIMES USED PER CLASS |
|---|---|---|
| Area | 699 | 58.3 |
| Congruent | 143 | 11.9 |
| Hypotenuse | 44 | 3.7 |
| Leg | 442 | 36.8 |
| Quadrilateral | 27 | 2.3 |
| Represent | 78 | 6.5 |
| Think | 246 | 20.5 |
| Triangle | 456 | 38.0 |
| Vertex | 20 | 1.7 |

*Source:* Chval, K. B., & Khisty, L. (2009). Latino students, writing, and mathematics: A case study of successful teaching and learning. In R. Barwell (Ed.), *Multilingualism in mathematics classrooms: Global perspectives.* Bristol, UK: Multilingual Matters. Used with permission.

## STOP AND THINK

Stop and think about what is noteworthy about Ms. Martínez's academic language use.

- As shown in Figure 8.4, Ms. Martínez said the word *leg*, referring to one of the short sides in a right triangle, 442 times in the first 12 mathematics lessons. What patterns would emerge in your academic language use if you were to record your first 12 lessons of the school year?

Let's peek into one of the lessons in which Ms. Martínez was found to use the word *leg* repeatedly. Transcript 8.6 takes place after the students have solved Problem 7 in Figure 8.5. Try It! 8.1 invites you to solve the problem before you read the transcript.

 **Try It! 8.1**

**Figure 8.5**   Problem 7 From Maneuvers With Triangles

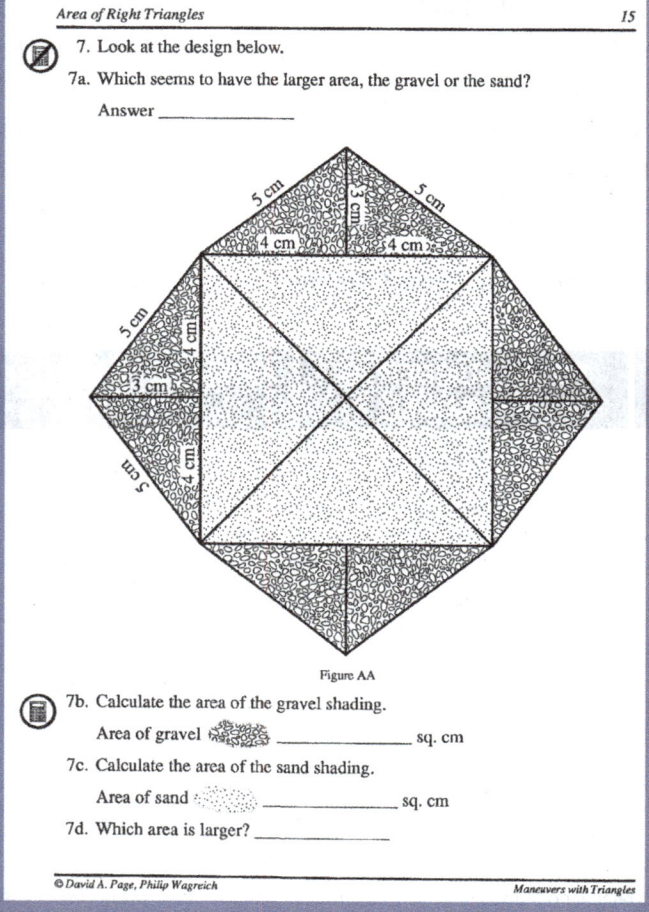

*Source:* Page, D.A., Wagreich, P., & Chval, K. (1993). *Maneuvers with triangles.* Palo Alto, CA: Dale Seymour. Reprinted with permission.

As you read Transcript 8.6 from the 12th day of school, consider how the term *leg* is used by Ms. Martínez and the multilingual learners in her classroom as they discuss finding the area of the gravel in the figure. Mariela is a student in her classroom. Ms. Martínez calls two girls to the chalkboard to explain their solution so that Mariela understands. This was a daily practice in Ms. Martínez's teaching. She wanted students to explain so that others understood their approaches. Explaining ideas to others was also an effective strategy for working out new ideas and locating areas where comprehension broke down.

**Transcript 8.6**

| | |
|---|---|
| Ms. Martínez: | Mariela wants to know how to find the area of the gravel [*referring to a shaded portion of Figure 8.5*]. So, if you give us your backs and you just write, Mariela will not understand what the heck it is you are doing. You need to talk. [*Ms. Martínez doesn't want Lorena and Miranda to just face the board and write their approach for solving the problem. She expects them to talk about their approach as they write it. She also wants them to look at the class, building their public speaking skills.*] |
| Lorena and Miranda: | Do we have to make a sketch? |
| Ms. Martínez: | If you want to make a sketch, of course, go ahead. |
| | [*Quiet for a long time*] |
| | That's a great-looking sketch, isn't it? [*Talking to students at their desks*] You are adding the *legs*? Why are you adding the *legs*? Because if you add the *legs* . . . Why are you adding the *legs*? Think about what we did a little while ago. [*Walking around the room*] Very nice. Very nice. While the girls are still writing, why don't you look at José's keystrokes and tell him what is wrong—give him a hint why this is not right? Or did you figure it out? |
| Ms. Martínez | [*talking to Rolando at his desk*]: Yeah, it sure is. I don't care what the answer is. Where did this answer come from? But where? Show me. This *leg*? And where did this other 4 come from? |
| Rolando: | [*No response*] |
| Ms. Martínez: | Where? It doesn't make any sense. You are going to have to pay very close attention to their explanation. |
| | [*Miranda and Lorena continue working at the board*] |
| Ms. Martínez: | Use your calculator. Where's your calculator? You've got a calculator, and you're not using it. [*Talking to students at desks as Miranda and Lorena write on the board*] Girls, are you almost there? |
| Miranda and Lorena: | Yes. |
| Ms. Martínez: | Where are the keystrokes [*referring to calculator keystrokes*] for the area of the square? How did you get . . . |
| Miranda and Lorena: | Okay, we are ready. |
| Ms. Martínez: | Yes, we are ready. We are going to listen now. Pay attention because you have to see if they are correct. |
| Miranda: | We got our triangle, our right triangle in there, we got to multiply one *leg* times the other *leg* to find the area of one triangle. |
| Ms. Martínez: | José, are you listening to that? |

(continued)

(continued)

| | |
|---|---|
| José: | Yes. |
| Ms. Martínez: | Lorena, can you stop and say what you've done so far? |
| Lorena: | Yes, we multiplied one *leg* times another *leg* on the triangle, and *N* equals 12. And the area is the whole rectangle, and we don't want the whole rectangle. We want the triangle. So, you have to divide by 2, [which] equals 6. Six is the area. Centimeters. Is the area. |
| Ms. Martínez: | I disagree. Anybody else disagree? |
| Class: | Yes. |
| Ms. Martínez: | You have to say that. When you see something you don't agree with, you say, "I disagree." If we don't disagree, then we don't have to discuss it. |
| Lorena: | Square centimeters [*a correction from* centimeters *to* square centimeters *in her response*]. |

Look back at Transcript 8.6 and consider:

▸ How did Ms. Martínez use the word *leg* in a mathematical context?

▸ How did Ms. Martínez's use of *leg* help Miranda and Lorena to begin to use it in their explanation?

In Transcript 8.6, you can see evidence of the way Ms. Martínez continued to use *leg* repeatedly with her students to facilitate their acquisition of it into their own language practices. You can also see how Miranda and Lorena appropriated this term in their explanation. Consequently, Ms. Martínez offers an illustration of how teachers can build meaning for mathematical language in context and over time where the focus is on meaning and mathematical processes, not on a single word or phrase (Chval & Khisty, 2009; Moschkovich, 2012, 2013, 2015).

## STRATEGIES FOR ENHANCING MULTILINGUAL LEARNERS' LANGUAGE DEVELOPMENT

Here is a list of strategies you can use to enhance multilingual learners' language development:

▸ Identify your students' familiarity with language. For example, if you are unsure if your students have familiarity, ask if they have heard or used the language, and ask for examples.

▸ Read aloud to students. This provides a context for discussing new or unfamiliar language (Manyak, 2012).

▸ Compare and contrast language (e.g., words, phrases) that has multiple meanings (Manyak, 2012). In this discussion, use the language in context and pronounce it clearly (Folse, 2008). You can also write it down on the board so that students recognize it.

▶ Let students use their first language as an anchor. Translation and use of cognates may enhance multilingual learners' language acquisition (Folse, 2008).

▶ Utilize students' everyday language to initially define or explain new language (Manyak, 2012; Moschkovich, 2013, 2015).

In addition to these strategies, Figure 8.6 includes a series of questions to support your decisions when introducing language.

**Figure 8.6** Language Decision-Making Table

| ASSESSING STUDENT UNDERSTANDING | BUILDING MEANING FOR VOCABULARY |
|---|---|
| You can ask your students the following guiding questions individually or in groups:<br><br>• What does this word/phrase mean to you?<br><br>• Have you used this word/phrase before?<br><br>• When have you used this word/phrase?<br><br>• What do you think of when you hear this word/phrase? Give me some examples.<br><br>• Can you draw or make a list of what you think of when you hear this word/phrase?<br><br>• What does this word/phrase mean in your first language? | You can use the following guiding questions to decide if you need to build meaning for language and how to do it:<br><br>• What are the most critical words/phrases that I should introduce/initiate (as the teacher)?<br><br>• Where is the most natural place for this language to be introduced?<br><br>• How can I build meaning through gestures, written words, images, videos, or models? Will it require role playing or a video?<br><br>• How can I include the language in future problems, activities, or thematic projects?<br><br>• How much time should I invest in building meaning for this language? |

# THINKING ABOUT ACADEMIC LANGUAGE IN YOUR PRACTICE

In this chapter, you explored two forms of specialized mathematical language: (1) language that has multiple meanings (i.e., everyday and mathematical meanings), such as *table* or *leftover*, and (2) language that has primarily a mathematical meaning, such as *quadrilateral, exponent,* or *right triangle.* To do this, you peeked inside the classrooms of Ms. Martínez and Ms. Bristow to identify strategies to facilitate multilingual learners' acquisition of academic language. Now, put these strategies into practice in Try It! 8.2.

**Try It! 8.2**

Select a lesson you will teach in the next month and identify what language may be new for multilingual learners. Then, use the following questions to guide your lesson planning.

| QUESTION | RESPONSE |
|---|---|
| 1. What language may have multiple meanings and uses, including mathematical meaning? | |
| 2. How important is it for all my students to know the different meanings associated with this language? | |
| 3. How will this language enhance multilingual learners' access to mathematical knowledge? | |
| 4. How can I build meaning for it? | |

## Reflect

- What are three different strategies you will use to facilitate multilingual learners' development of specialized mathematical language in your classroom?

# CHAPTER 9
## USE YOUR DISCOURSE STRATEGICALLY TO ENHANCE MULTILINGUAL LEARNERS' OPPORTUNITIES TO LEARN

### Key Concepts

In this chapter, you will

- ✓ explore the different kinds of discourses you use.

- ✓ discuss the importance of teacher discourse to facilitate learning.

- ✓ discuss instructional strategies for enhancing discourse in the classroom.

In the previous chapter, we discussed academic language. Typically, academic language is thought of as particular lexical items such as vocabulary words. This is true, but when we examine our teaching, academic language in a broader sense becomes an important element in providing equitable instruction for multilingual learners. One way to signal this broader focus is to use the word *discourse*. Defined as written or spoken communication, discourse is usually thought of as language that extends beyond single words or short phrases to take into account lengthier spoken or written communication.

Spoken discourse plays a vital role in teaching as it is often the central means of conveying information and evaluating students' ideas in classroom discussions. Written discourse—both the teacher's use of written communication and students' written discourse—is also used in the classroom. In this chapter, we will focus on both spoken and written discourse involved in teaching since written discourse is an important element for providing an anchor for spoken discourse during instruction, as well as for building academic literacies.

In considering the role of discourse in classroom teaching, it can be helpful to reflect on our encounters with people in different contexts. In these instances, we recognize different dialects, meanings, and communication approaches. These interactions always fascinate us and inspire our awe for language. For example, consider how you would change your word selection, voice, tone, volume, and body language in the following contexts:

▶ You are explaining a concept to a baby or toddler.

▶ You are convincing your partner or friend to relocate.

▶ You are chatting or texting with your best friend online.

▶ You are giving a presentation to colleagues or to the local school board.

▶ You are teaching a mathematics lesson to your class.

After spending considerable time in classrooms, we noticed instances when language was not accessible to multilingual learners. For example, a teacher used expressions, such as "he walked five blocks," or idioms, such as "knock your socks off," that were unfamiliar to some students. We also noticed that some teachers were more adept at facilitating the development of students' language and mathematical knowledge. These expert teachers were attuned to the ways they could leverage their *discourse*, or specific use of language, to create mathematical learning and language acquisition opportunities. "Teachers must actively work to make the discourse environment of the classroom welcoming to and respectful of students from all cultural backgrounds if the classroom is to afford learning opportunities for all children" (Cazden & Beck, 2003, p. 170). Moreover, for students to acquire the language of mathematical discourse (e.g., referencing concepts, justifying their solution, or explaining a particular method), teachers must model such behavior (Huang, Normandia, & Greer, 2005; Schleppegrell, 2004). Teachers are responsible for promoting specific kinds of discourse for students. Therefore, the discourse a teacher uses can enhance or constrain multilingual learners' opportunities to learn, and teachers can purposefully change their discourse to facilitate learning. Based on these situations, we began to ask: How can teachers use and discuss language so that multilingual learners are successful in mathematics classrooms? What are specific things that teachers say that facilitate opportunities to learn for multilingual learners? What are strategies that can enhance teacher discourse?

*The discourse a teacher uses can enhance or constrain multilingual learners' opportunities to learn, and teachers can purposefully change their discourse to facilitate learning.*

## REFLECTING ON VARIETIES OF ENGLISH THAT INFLUENCE DISCOURSE

English currently serves as a common international language with many countries in the world using English, in addition to their native language(s), for purposes related to education, science, medicine, and business. There are many benefits to having a common international language. However, the colonizing history and effects of English in many countries must also be noted, as the detriment to native languages and cultures globally cannot be underestimated (Baker & Wright, 2017). Due to this history, there are varieties of English in use internationally, sometimes referred to as World Englishes, that reflect the intersection of particular types of English with different cultures and languages. The varieties of English that have evolved globally mean that multilingual students in U.S. schools frequently come from countries and continents such as China, India, and parts of Africa, where British English is often the norm. In settings such as these, English words, phrases, slang, and idioms are different from those used in U.S. English. For instance, while dining in a restaurant in the United States, a person who orders "chips" with a sandwich will get potato chips. However, in the United Kingdom, when "chips" are ordered, french fries will be on the plate with the sandwich. A biscuit in the United States is a type of bread, while in the United Kingdom it refers to a cookie.

Alternatively, when Kathryn's children were young, they enjoyed reading the different (i.e., British and U.S.) versions of *Harry Potter* by J. K. Rowling. They would often compare how the language differed. For example:

> **"** He was in a very good mood until lunch-time, when he thought he'd stretch his legs and walk across the road to buy himself a bun from the *baker's opposite*. (Rowling, 2017a, p. 9, British version, emphasis added)
>
> He was in a very good mood until lunch-time, when he thought he'd stretch his legs and walk across the road to buy himself a bun from the *bakery*. (Rowling, 2017b, p. 4, U.S. version, emphasis added) **"**

"Baker's opposite" means a bakery that is opposite/across from his house.

Linguists identify three major dialects of English: British Isles, North America, and Australasia. In addition, there are differences in regional dialects across the United States. Video 9.1, "How Do You Pronounce 'Water'?" (*Washington Post*, 2014), illustrates some of these differences. If you would

**VIDEO 9.1:**

*Washington Post.* (2014, July 8). *How do you pronounce "water"?* [Video]. https://www .washingtonpost.com/ video/national/how-do-you-pronounce-water /2014/07/09/3ef471c2-0760-11e4-9ae6-0519a2bd5dfa_video.html

**VIDEO 9.2:**

David Pakman Show.
(2015, June 27). *The rapidly
changing language of
American English* [Video].
YouTube. https://youtu
.be/4YSbNaXaOy0

like to know more about the history of variations in English pronunciations across the United States, William Labov, a linguist, discusses the shift in English pronunciation pre– and post–World War II and how it has influenced present-day pronunciations in Video 9.2, "The Rapidly Changing Language of American English" (David Pakman Show, 2015). Further, U.S. English has many variations across geography, depending on speakers' cultural background, profession, social class, and age, among other factors.

While teaching at our universities, we began to notice expressions, idioms, and slang that we use that were unfamiliar to our students from international settings. Our international colleagues would politely ask, "What does that expression mean? What is the origin of that expression?" As we worked in elementary classrooms, we noticed slang that teachers and students used that was likely unfamiliar to multilingual learners. However, the multilingual learners did not typically ask teachers to clarify meanings. Examine common slang expressions from the 1960s and 2020s. Do you know the meanings of all of them?

| 1960s | 2020s |
|---|---|
| It's a gas | He's so extra |
| That's boss | Big yikes |
| Pedal pushers | Cap/no cap |
| Drop a dime | Throw shade |
| Going steady | That party was lit |
| Padiddle | You slay me |

The meanings are provided in Appendix B. We note these ideas because if we use idioms or cultural expressions that multilingual learners do not understand, we can be inadvertently positioning these students as outsiders during classroom discussions. Creating an inclusive classroom compels us to consider the language we use and not assume that every learner understands what we say.

## STOP AND THINK

Stop and think about the expressions, idioms, and slang that you use.

- How often have you asked your students what those terms mean?

- How often do you ask your students what particular terms they use mean?

## WHAT THE RESEARCH SAYS ABOUT DISCOURSE

According to linguist James P. Gee (1999), "If I had to single out a primary function of human language, it would be not one, but the following two: to scaffold the performance of social activities (whether play or work or both) and to scaffold human affiliation within cultures and social groups and institutions" (p. 1). Being a teacher, a mother, a soccer player, or an employee are identities that mark affiliations to social groups; such identities are largely defined by discourse. Thus, *discourse* is not just a linguistic ability, but also expresses our affiliations to specific social groups (J. P. Gee, 2015). For instance, your discourse illustrates your age, geographical region, profession, social class, ethnicity, and association to subcultures (e.g., the specific discourse used by a group of bikers or the fans of *Star Trek*). According to the context, you emphasize or employ some discourses over others.

Bakhtin (1986) called a "social language" those variations of language that change according to the social context. In other words, social language can be thought of as a "verbal resource we associate with, say, being a mathematician" (Khisty & Chval, 2002, p. 156). This idea can be extended to include the specific discourse a teacher uses in the classroom and the beliefs, values, and identities contained in it. Figure 9.1 contains a summary of key ideas from linguistic research (J. Gee & Gee, 2007; Jenkins, 2014; Labov, 2011; Rymes, 2014, 2016).

**Figure 9.1**  Ideas About Discourse From Linguistic Research

| SOME IDEAS ABOUT DISCOURSE | EXAMPLES |
|---|---|
| There is not "one" English, but multiple variations of English. These different kinds of English combined with our identities, beliefs, and values are called *discourses*. | English varies according to geographical location. For example, specific English expressions used in the southern United States vary from those used in the northeastern United States. |
| Certain variations of language are more appropriate to specific contexts and circumstances (e.g., in professional contexts, appropriate jargon is necessary to express precise meanings; however, the same jargon may not be suitable in other social contexts such an informal dinner with friends). | Professional groups have different dialects. For example, an engineer talks differently than a physician or teacher. Language also varies by age and subculture (quilting group or crew team). |
| Even at the individual level, we change our discourses depending on the context and our purposes. We use certain variations of English depending on the circumstance. | The same person greets:<br>• Hey, dude!<br>• Good morning, Dr. Thompson.<br>• Hi, Mom.<br>• Hello, sweetie. |

Teachers have the capability to enhance and transform student learning through their discourse. When teachers decisively introduce discourse that is filled with rich language, students appropriate that discourse as their own, use it as a tool for their thinking, and use it as a tool to communicate their thinking (Khisty & Chval, 2002).

## TEACHER DISCOURSE

Teachers use specialized discourse that may not be used or experienced in other contexts. In the following sections, you will consider how teachers

▶ connect academic language to mathematical representations,

▶ pose questions to engage multilingual learners, and

▶ problematize content.

## Connecting Academic Language to Mathematical Representations

Within mathematics, it is critical that students be able to effectively coordinate within and across different mathematical representations (e.g., between the term *remainder* and *R* or between the term *picture graph* and a picture graph). However, for students to develop this ability, they must be given opportunities to make connections between representations and academic language—specifically specialized mathematical discourse.

## STOP AND THINK

Stop and think about the last time you introduced a new mathematical representation.

- What steps did you take to facilitate students' understanding of the specialized mathematical language used for it and the representation itself?

Recall the way Ms. Bristow introduced the term *remainder* to her students in Chapter 8 by leveraging their informal language and prior experiences (i.e., *leftovers*). In Transcript 9.1 (Pinnow & Chval, 2014), the lesson has continued, and Ms. Bristow introduces students to the convention of writing an *R* to represent the remainder in division problems.

**Transcript 9.1**

| | |
|---|---|
| Ms. Bristow: | [*Stands in front of the class at the SMART Board®. Students are seated in front of her on the carpet.*] Usually, when you are dividing something, you want to divide it fairly so everybody gets the same amount. Because that will usually cause problems if everybody does not get the same amount. Have you guys noticed that that causes problems in your family? Yeah, and it causes problems at school, too. So, everybody wants the same amount, and sometimes there are leftovers or remainders to share something. So, we are going to have some situations that are all about balloons. Because we know we can't cut up the balloons because they wouldn't then be balloons. [*Displays the handout that students will use in their pair work on the SMART Board®. The problems on the handout involve names of children in the classroom.*] . . . <br> So, this one says, "Ms. Bristow had 18 balloons at her party. She wanted to give 3 people balloons. How many balloons can she give each person?" That is one type of division problem. Remember we are trying to figure out how many in each group for division problems. So, then there is this other part here. This says, "Casey had 17 balloons to give to friends. She wanted to give 5 balloons away to each person. How many people can she |

give balloons to?" So, these are the two different types of problems for division that we talked about [*referring to partitive and measurement division situations introduced in earlier lessons*]. So, sometimes you know how many are in each group. Like, Casey knows she is giving 5 balloons away, but she doesn't know how many people. Over here, we know how many groups there are, but we don't know how many go in every group. So, those are two types of situations. When you write a problem that has a remainder, do you see how I have got this number sentence here? [*Ms. Bristow's handout includes: Number Sentence* _____ ÷ _____ = _____ *Remainder:* _____] And I have this extra word here: *remainder* [*elongates the vowel in the word* extra *and speaks slightly louder to emphasize the importance of this point*]. Sometimes people will just shorten it [*writes an* R *above the word* remainder] and write an *R* there. But today I have written that whole word, *remainder*.

Lauren:          So, we know what that *R* means.

Ms. Bristow:     Yeah, so you know what that *R* means. Otherwise, it wouldn't make much sense, would it?

Lauren:          Yeah.

*Source:* Pinnow and Chval (2014).

## STOP AND THINK

Stop and think about how Ms. Bristow introduced a new mathematical representation.

- How did Ms. Bristow connect students' language with the mathematical representation of *R* for remainder?

Ms. Bristow introduces the handout that students will use with their partners. In this handout, Ms. Bristow uses a number sentence involving remainders to discuss the symbol for remainder (i.e., the capital letter *R*) used to report answers for division situations at the third-grade level. Ms. Bristow draws students' attention to the symbol by posing a question to get them to notice this form ("Do you see how I have got this number sentence here? And I have this extra word here: *remainder*"). Ms. Bristow elongates her tone on the word *extra* and speaks slightly louder to draw students' attention to this form. She recognizes that students have written number sentences during their elementary years, but they have not yet written number sentences involving remainders. She then writes the letter *R* above the word *remainder* to show students how this symbol relates to the term *remainder* in number sentences. By orchestrating discussion about the symbolic use of the letter *R*, slowing her speech, stressing important words and ideas, and writing the symbol as she explains the use of it, Ms. Bristow created scaffolded discourse that multilingual learners need to make connections to the academic uses of language in this context. This is important because directly after this discussion, students will be paired up to solve mathematics problems and will encounter and use this symbol.

## Posing Questions to Engage Multilingual Learners

Teachers ask students a number of questions during the course of a mathematics lesson, such as "Why did you decide to do ____?" or "How did you solve this problem?" Questions serve a range of purposes, including eliciting and extending student thinking. Prior research has found that questions used to probe students' mathematical explanations can foster mathematical thinking (Franke, Webb, Ing, Chan, & Freund, 2007; Herbel-Eisenmann, Steele, & Cirillo, 2013; Webb et al., 2009; Webb et al., 2008) and increase student engagement (Cazden & Beck, 2003). Moreover, when such questions are used with multilingual learners, their opportunities to use the target language are enhanced. Yet, some teachers remain hesitant to ask such questions of multilingual learners and may rely more heavily on simple or closed questions that demand little from learners and provide little opportunity to use extended discourse as they respond (e.g., questions that only require a yes/no answer). This is an important point because when multilingual learners are not provided opportunities to answer more challenging questions in class, this can position them as incompetent to do so. Students notice which of their peers are asked more difficult questions consistently and which ones are posed simple, closed-ended questions repeatedly. As the research indicates (Pinnow & Chval, 2015; Yoon, 2008), students base their treatment of multilingual learners in the classroom on how the teacher has positioned those learners. As we noted in prior chapters, multilingual learners also notice when their teachers seem to expect little of them. Thus, it becomes vital that teachers create a high-challenge, high-support classroom environment for multilingual learners (Gibbons, 2009).

*When multilingual learners are not provided opportunities to answer more challenging questions in class, this can position them as incompetent to do so.*

Ms. Martínez regularly asked her multilingual learners challenging questions. In Transcript 9.2 (adapted from Khisty & Chval, 2002), you'll look at one example. Transcript 9.2 is broken down into four columns. Therefore, to read the transcript in order, read the numbers in sequence. This format highlights when Ms. Martínez asks questions and *revoices* student responses. Teachers can strategically use revoicing to contrast or align students' ideas as a means to engage students in mathematical discussion (Cazden & Beck, 2003; Herbel-Eisenmann et al., 2013).

*Revoice: To restate or rephrase language said by another person*

Transcript 9.2 begins with a problem involving a rectangle. Near the beginning of the year, the students were given two problems. The first problem involved a rectangle with an area of 600 square meters and a width of 30 meters. On the chalkboard was a drawing of a rectangle with an area of 600 square meters marked inside of it and 30 meters marked on one side. The students were asked to find the missing length. The second problem involved a rectangle with a perimeter of 2 meters and a length of $\frac{3}{4}$ meter also with a drawing, and the students were asked to find the missing width. The students have finished solving the two problems, and now Ms. Martínez asks for a pair to volunteer to solve one of the problems on the board. Two girls are selected, and as they go to the board, they are reminded to speak loudly as they "talk through every step." As you read the transcript:

▶ What do you notice about Ms. Martínez's questions?

▶ How did Ms. Martínez use questions to facilitate her students' language development and mathematical learning?

## Transcript 9.2

To read this transcript, follow the numbers in order.
S = Student
T = Teacher

| STUDENT | TEACHER REVOICING/ RESTATING | TEACHER QUESTIONING | TEACHER DIRECTING STUDENTS |
|---|---|---|---|
| 1 S1: We divide the 600 by 30. | 2 T: You divided the 600 square units by 30. | 3 T: Why did you divide the area by the side length? | |
| 4 S1: To find the side. | 5 T: To find the side length. | 6 T: Now, will that work? Can you give me a different kind of problem to show me why it will work? Can anybody show me a different problem so I can know why it works? | |
| | | | 7 T: Okay, Rocky. [*Another student, Rocky, joins the others at the board.*] |
| 8 S2: With a different square [*Rocky draws a rectangle and labels the corresponding parts 6 square meters and 3 meters*]. | | 9 T: How would you find the area of that square, really, rectangle? | |
| 10 S2: Work it out. | | 11 T: How are we going to find the side length? What's the idea behind doing what we're doing? | |
| 12 S2: You're going to divide. | | 13 T: Why are you going to divide, Rocky? Will you explain? | |
| 14 S2: Six divided by the 3. | | 15 T: Why is it going to work? What's the relationship—like that between multiplication and division? Are they related? | |
| 16 Choral: Opposite. | 17 T: One is the inverse of the other, the opposite of the other. | 18 T: So if the area is multiplying, then what is the opposite of multiplication? | |
| 19 Choral: Dividing. | 20 T: Dividing. So that's why this would work. | | |

*Source:* Adapted from Khisty and Chval (2002).

As you read Transcript 9.2, you may have noticed the sheer number of questions Ms. Martínez asked her students during the discussion. In Try It! 9.1, look more closely at her questions and write the underlying purpose of each in the right-hand column.

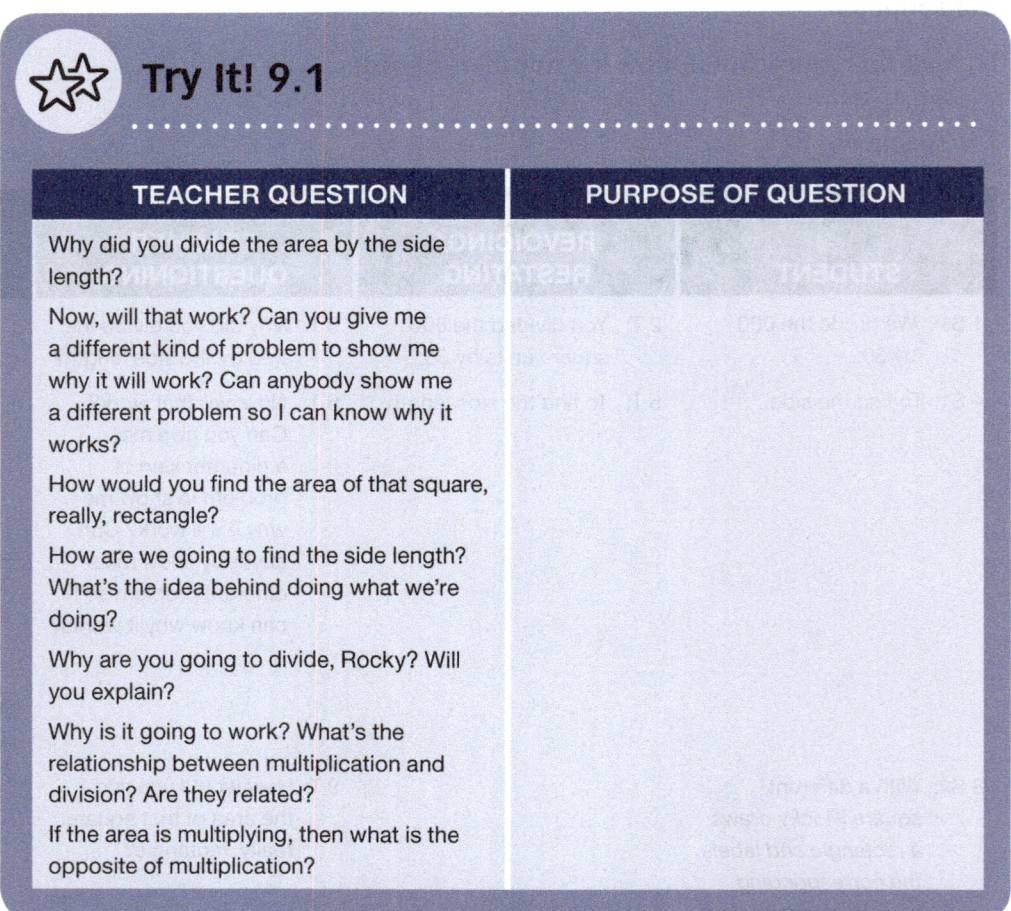

### Try It! 9.1

| TEACHER QUESTION | PURPOSE OF QUESTION |
| --- | --- |
| Why did you divide the area by the side length? | |
| Now, will that work? Can you give me a different kind of problem to show me why it will work? Can anybody show me a different problem so I can know why it works? | |
| How would you find the area of that square, really, rectangle? | |
| How are we going to find the side length? What's the idea behind doing what we're doing? | |
| Why are you going to divide, Rocky? Will you explain? | |
| Why is it going to work? What's the relationship between multiplication and division? Are they related? | |
| If the area is multiplying, then what is the opposite of multiplication? | |

*Source:* Adapted from Khisty and Chval (2002).

Ms. Martínez's questions serve a range of purposes, such as clarifying students' mathematical thinking and making connections to inverse operations. Across the questions, it's important to recognize the ways Ms. Martínez continually sets high expectations for her students as evidenced in her discourse. She also used lengthier discourse, not just single words or vocabulary items, to extend student spoken discourse. This focus on lengthier discourse created a platform for encouraging students' critical thinking and mathematical problem solving.

## Problematizing Content

For students to learn, value, and appreciate mathematics, teachers must establish a need or purpose for it (Ball & Forzani, 2011). To do this, teachers use their discourse to introduce mathematical tasks and content in different ways, depending on the topic. Ms. Martínez and Ms. Bristow often introduced tasks by suggesting they were experiencing problems and they wanted the students to help them in those situations. For example, Ms. Martínez wanted to help a friend, or Ms. Bristow was organizing a fundraiser.

## STOP AND THINK

Stop and think about the different ways you introduce mathematical problems or tasks to your students.

- How do you create a need or purpose for mathematical topics, tasks, or problems?

- How do you problematize situations?

- How would you introduce "Mystery Coin" problems such as "Your mystery bag has 43 cents. What coins could you have in your bag?" to your third-grade students?

Prior to introducing the "Mystery Coin" problems (from Chval & Pinnow, 2018) to her class, Ms. Bristow wanted to ensure her multilingual students and other students had sufficient familiarity with different kinds of U.S. coins. Therefore, Ms. Bristow developed an initial lesson that investigated actual U.S. coins through touch and provided time for all students to learn about the shape, size, and value of each coin and communicate about what they learned in speech and writing. This innovative approach was engaging for all of her students in the class, even those with experience with U.S. coins. She provided each student with a brown bag with a penny, nickel, dime, and quarter in it. Each student reached inside their own bag to identify, through touch alone, various features of each coin, and then wrote descriptions about the coins. This novel approach fostered meaningful engagement of all the students, regardless of their experience with U.S. coins. Mateo's descriptions of the coins are shown in Figure 9.2.

**Figure 9.2** Mateo's Responses About Coins

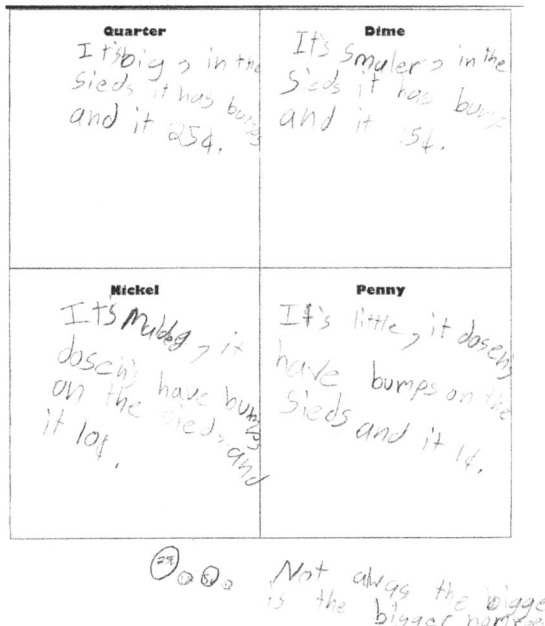

After students had experience with the coins, Ms. Bristow then moved to create a need to effectively and efficiently count them. In Try It! 9.2, examine how Ms. Bristow used her discourse to problematize the situation ("When I count coins . . . I wasn't sure where to start") and provided opportunities for students to use mathematical discourse themselves. In the left-hand column, you'll see the transcript from the classroom discussion. In the right-hand column, notate the purpose of Ms. Bristow's discourse.

 **Try It! 9.2**

Identify the ways Ms. Bristow established this need in the transcript and write it in the right-hand column of the table.

T = Ms. Bristow

Ch = Choral response

| TRANSCRIPT | | PURPOSE OF MS. BRISTOW'S DISCOURSE |
|---|---|---|
| T: | So, yesterday we just kind of looked at the coins, but we didn't count the coins. | Refers to previous lesson to build on prior knowledge. Distinguishes looking at coins from counting coins. |
| | When I count coins . . . I wasn't sure where to start. | Establishes a problem that she needs help solving. In other words, she problematizes the content. |
| | Um, because I know a penny—the value of a penny is one cent—and so sometimes I think, well, we start with one all the time when we're counting. We'll start like "one, two, three, four," so we start with one a lot. | |
| | Do you think starting to count with a penny would be a good strategy for me to count the coins? | |
| Ch: | No. | |
| T: | Why is that not a good strategy for me, 'cause I count with ones all the time? What do you think, Samantha? | |
| Samantha: | It takes, like, forever. | |
| T: | What do you mean it takes forever? | |
| Samantha: | Because if you go up one all the way to a hundred that will waste up your time. | |
| T: | Okay, but I don't have a hundred, I have—I don't have a lot of pennies. I only have one penny. | |
| | Why can't I count this just—I'll start with the penny. | |
| Samantha: | Because it might be a little bit harder. You could maybe do like [*moves coins in order of value*]. | |
| T: | Oh, so I could just do biggest to—I could do biggest to smallest. So, I could—or I could start with one. | |

| | |
|---|---|
| Samantha: | No, don't start with the smallest number; start with the biggest number. |
| T: | Biggest—like here's the biggest coin [*picks up quarter*]. |
| Samantha: | No, the biggest number, like 25, 10, 5. |
| T: | Why? |
| | Samantha's saying, "Okay, I'm not going to start with the biggest coin" [*referring to the size*]. |
| | I'm not going to go quarter, then count the nickels, then count the pennies, then count the dimes. |
| | Yasmin [*shaking head no*] says, "No, you shouldn't do that." So, do you agree with Samantha then? |
| Yasmin: | Yeah. |
| T: | So, she's saying not the largest size, but the largest value is the easiest way to start, with like 25. |
| | Why is it easier for me to count starting with a 25 or a quarter? |

Look back at the transcript (Chval & Pinnow, 2018) and identify how Ms. Bristow motivated the lesson. Compare your responses with our thoughts in Figure 9.3.

**Figure 9.3** Purpose of Ms. Bristow's Discourse Revisited

T = Ms. Bristow

Ch = Choral response

| TRANSCRIPT | PURPOSE OF MS. BRISTOW'S DISCOURSE |
|---|---|
| 1. T: So, yesterday we just kind of looked at the coins, but we didn't count the coins. | 1. Refers to previous lesson to build on prior knowledge. Distinguishes looking at coins with counting coins. |
| 2. When I count coins . . . I wasn't sure where to start. | 2. Establishes a problem that she needs help solving. In other words, she problematizes the content. |
| 3. Um, because I know a penny—the value of a penny is one cent— | |
| 4. and so sometimes I think, well, we start with one all the time when we're counting. We'll start like "one, two, three, four," so we start with one a lot. | 4. Identifies previous common practice to contrast it with a new counting approach. |
| 5. Do you think starting to count with a penny would be a good strategy for me to count the coins? | 5. Uses a yes/no question to ensure students do not utilize known counting procedures. |
| 6. Ch: No. | |
| 7. T: Why is that not a good strategy for me, 'cause I count with ones all the time? What do you think, Samantha? | 7. Poses a follow-up question to establish why counting by ones would be problematic. This is the second time she owns this strategy with the words "for me." She also refers to the strategy as "good" to emphasize that starting with the penny could be done to obtain the correct answer; however, it may not be the most efficient. |
| 8. Samantha: It takes, like, forever. | |
| 9. T: What do you mean it takes forever? | 9. Probes Samantha for clarification. |

(continued)

(continued)

| TRANSCRIPT | PURPOSE OF MS. BRISTOW'S DISCOURSE |
|---|---|
| 10. Samantha: Because if you go up one all the way to a hundred that will waste up your time. | |
| 11. T: Okay, but I don't have a hundred, I have—I don't have a lot of pennies. I only have one penny. | 11. Restates situation and how it differs from the situation posed by Samantha. |
| 12. Why can't I count this just—I'll start with the penny. | 12. Restates question under consideration. |
| 13. Samantha: Because it might be a little bit harder. You could maybe do like [*moves coins in order of value*]. | |
| 14. T: Oh, so I could just do biggest to—I could do biggest to smallest. So, I could—or I could start with one. | 14. Restates initial idea under consideration (of starting with a penny). |
| 15. Samantha: No, don't start with the smallest number; start with the biggest number. | |
| 16. T: Biggest—like here's the biggest coin [*picks up quarter*]. | |
| 17. Samantha: No the biggest number, like 25, 10, 5. | |
| 18. T: Why? | |
| 19. Samantha's saying, "Okay, I'm not going to start with the biggest coin" [*referring to the size*]. | 19. Asks class to consider why Samantha's idea might work. |
| 20. I'm not going to go quarter, then count the nickels, then count the pennies, then count the dimes. | 20. Provides example of a counting combination starting with the largest diameter to the smallest diameter. |
| 21. Yasmin [*shaking head no*] says "No, you shouldn't do that." So, do you agree with Samantha then? | 21. Pulls another student into discussion and connects her thinking with Samantha's. |
| 22. Yasmin: Yeah. | |
| 23. T: So, she's saying not the largest size, but the largest value is the easiest way to start, with like 25. | 23. States Samantha's idea again to count by the value rather than the size of the coin. |
| 24. Why is it easier for me to count starting with a 25 or a quarter? | 24. Prompts class to consider why Samantha's strategy is easier. |

*Source:* Chval and Pinnow (2018).

Before Ms. Bristow introduces "Mystery Coin" problems, she needs to assess if students are able to count the value of different combinations of coins and count them efficiently (e.g., starting with the largest value, which differs from previous counting experiences that start with 1). In the next lesson, after Ms. Bristow was sure that all students were familiar with the U.S. coins, their values, and counting values of coins, she introduced "Mystery Coin" problems as follows: "Today we're going to still have mystery bags, but we're going to have some problems to work out with these bags. Now I want you to imagine that inside one of our bags we have different amounts of coins." She referenced back to the previous lesson where the brown bags had one of each coin.

> *Now imagine that in our bags we don't know the coins. I don't know the coins I have in my bag, but I know they're worth a certain amount of money. So, I know that there's 54 cents in my bag [writes "54¢" on easel]. I have 54 cents in my bag, so . . . what can I do, how can I figure out what coins could be in my bag? So, I'm not—I can't look, I can't peek in, and I'm not going to feel around; I'm going to keep it closed [the bag]. How could I figure out what coins could be in this bag? Any ideas?*

Ms. Bristow's purposeful use of discourse engaged her students and provided a foundation for their success with solving the coin problems.

# STRATEGIES FOR PROMOTING CLASSROOM DISCOURSE

Mercer (1995, p. 32) identified the following strategies that promote and initiate classroom discourse:

▶ Make a declarative (open-ended or provocative) statement that invites a rejoinder or disagreement;

▶ Invite elaboration ("Could you say a bit more about that?");

▶ Admit perplexity when it occurs, whether about the topic itself or about a pupil's contribution to it;

▶ Encourage questions from pupils (rare in many classrooms); and

▶ Maintain silence at strategic points (Dillon [(1982), another classroom researcher,] suggests that three to five seconds may be enough to draw in another pupil's contribution or encourage the previous speaker to elaborate on what was said).

More recently, de Araujo, Roberts, Willey, and Zahner (2018) reviewed research and identified four broad categories of practices that support multilingual learners to develop or access mathematics discourse. Those four categories—(a) eliciting, (b) modeling, (c) revoicing, and (d) recognizing and valuing multilingual learners' multiple resources—are summarized in Figure 9.4.

**Figure 9.4**　Strategies From a Review of Research

| STRATEGY | IDEAS FROM DE ARAUJO ET AL. (2018, PP. 902–903) |
|---|---|
| Eliciting | Eliciting discourse draws students into the verbal interactions" (Hansen-Thomas, 2009, p. 94). |
| | Hansen-Thomas (2009) found a successful teacher used choral responses, modeling standard forms, and encouraging the use of mathematical language in small and large groups. |
| | Honor students' lived experiences and be cognizant of their evolving identities (Celedón-Pattichis & Turner, 2012; Takeuchi, 2015; Turner, Dominguez, Empson, & Maldonado, 2013; Turner, Dominguez, Maldonado, & Empson, 2013; Warren & Young, 2008). |
| | Recognize the resources students bring to the classroom and affirm students' identities. |
| Modeling | Model discourse directly (Khisty & Chval, 2002) or through structured reading of the text, calculating, or solving problems orally and repeating or emphasizing algorithms, concepts, formulas, and definitions (Hansen-Thomas, 2009). |
| Revoicing | Use revoicing to "positively acknowledge or redirect students to correct forms" (Hansen-Thomas, 2009, p. 94). Revoicing involves "explicit verbal, gestural, and other non-verbal positioning moves by the teacher that . . . explicitly plac[e] the original speaker in relation to other people, the task, or the original speaker's interpretation of his or her own utterance" (Enyedy et al., 2008, p. 141; Herbel-Eisenmann et al., 2013). |
| Recognizing and Valuing Multilingual Learners' Multiple Responses | Draw on students' first language as a tool to help multilingual learners comprehend oral or text-based mathematics discourse (Enyedy et al., 2008; Hansen-Thomas, 2009; Khisty & Chval, 2002; Tavares, 2015), allow students to code switch (Salehmohamed & Rowland, 2014; Setati & Adler, 2000), and use students' funds of knowledge (Lipka, Sharp, Adams, & Sharp, 2007). |

## THINKING ABOUT YOUR DISCOURSE IN YOUR PRACTICE

In this chapter, you identified differences in discourse, different dialects, and different interpretations for the same expressions (e.g., "chips"). In addition, you peeked into the classrooms of Ms. Martínez and Ms. Bristow to examine how they connected academic language to mathematical representations, posed questions to engage multilingual learners, and problematized content. Multilingual learners in both of these classrooms appropriated the language that their teachers used. Since teachers spend a significant amount of time with their students, they have the ability to use discourse that enhances multilingual learners' opportunities to learn language and mathematics.

> *Since teachers spend a significant amount of time with their students, they have the ability to use discourse that enhances multilingual learners' opportunities to learn language and mathematics.*

## STOP AND THINK

Stop and think about your use of discourse.

- If you videotaped your classroom tomorrow, what would you notice about your discourse?

Teacher beliefs about mathematics and acceptable ways of communicating mathematically are relayed through their discourse. The ways that teachers focus attention on strategies impact which mathematical practices are valued (Wood, 1998). For example, when teachers use students' problem-solving strategies to draw attention to mathematical ideas, the value of mathematical thinking is relayed. In contrast, when teachers use strategies to draw attention to one selected way of solving the problem, the value of finding this preselected method is enforced. Similarly, when teachers place emphasis only on correct answers, the value of learning from and through failure is lost.

## Reflect

- How does your discourse influence multilingual learners' mathematical practices in your classroom?

- What changes would you like to make to your discourse within the next year?

# CHAPTER 10
## FOSTER A CULTURE OF WRITING IN THE MATHEMATICS CLASSROOM

### Key Concepts

In this chapter, you will

- ✓ explore ways to establish a classroom environment that values mathematical writing.

- ✓ identify ways to foster a culture of writing in the mathematics classroom.

Writing is a social practice and a competence that is often overlooked in the mathematics classroom. However, writing plays an important role in developing mathematical thinking and, therefore, should be part of multilingual learners' instruction. Through writing, you can assess students' understanding and development of concepts (Brozo & Crain, 2018). Writing can also help students develop mathematical reasoning and consolidate their thinking as they reflect on their work (Kosko & Zimmerman, 2019; National Council of Teachers of Mathematics, 2000). In our conversations with teachers, many have voiced hesitation in focusing on writing in the mathematics classroom because they weren't sure how to effectively foster mathematical writing. Oftentimes, these teachers did not have prior education or professional development on mathematical writing and weren't sure what they should or could do differently. This led us to develop this chapter.

> *Writing plays an important role in developing mathematical thinking and, therefore, should be part of multilingual learners' instruction.*

## REFLECTING ON YOUR EXPERIENCES

In Try It! 10.1, you are asked to write, then reflect on and revise, a paragraph explaining to parents what students will be learning in your mathematics classroom.

 **Try It! 10.1**

Write a paragraph for a parent newsletter that explains the mathematical ideas students will learn in the next month or semester. Include an activity parents can do with their children to facilitate their mathematical learning. Be sure to highlight how the activity will complement what students will do in the classroom. Then read your paragraph and revise based on the following questions:

- Where can you provide more clarity?

- Did you consider the audience of the text? If not, how could the text be revised for the audience?

Ask a colleague or friend to read the paragraph and provide feedback.

 **STOP AND THINK**

Stop and think about your own writing processes.

- What are some things you consider as you write?

- What are your ideal writing conditions?

- What types of feedback are most helpful for your writing? Why?

## WHAT THE RESEARCH SAYS ABOUT WRITING IN MATHEMATICS

Writing is an integral part of mathematics and second language learning. For that reason, mathematical writing should be used within a model of integrated content learning and language development (Chval & Khisty, 2009). It is important to note that writing is one of the most difficult language domains (i.e., speaking, reading, listening, and writing) for multilingual learners to develop. In fact, language domains do not develop in the same way or at the same rate (Collier, 1995; Cummins, 2008). Multilingual learners in your classroom may demonstrate proficiencies in other language domains before becoming proficient in writing. Although writing in mathematics, as opposed to other content areas, can create additional challenges for students due to its complexity, it should be fostered and developed in order to increase the students' mathematical thinking abilities. Therefore, developing a culture that appreciates and values writing in the mathematics classroom is critical to multilingual learners' development.

The teaching of writing must be more than the acquisition of motor skills, such as good handwriting, and non-motor skills, such as grammar and spelling. It should be "cultivated" in a manner that arouses a need to communicate mathematical ideas clearly to an audience while simultaneously creating a culture for writing in the classroom (Vygotsky, 1978). A culture of writing includes the beliefs about writing, the ways students value writing, and the social practices shared around writing in your mathematics classroom. In this chapter, we explore strategies to foster a community that facilitates a culture of writing in the mathematics classroom with multilingual learners.

### STOP AND THINK

Stop and think about the ways you cultivate a culture of writing in your mathematics classroom.

- What are some strategies you currently use?

## ESTABLISHING A NEED TO WRITE IN MATHEMATICS

A critical aspect of a culture of writing is creating a need for mathematical writing. Although you can provide several different kinds of mathematical writing activities, this is not enough to create a need. Writing should have a purpose and be incorporated in authentic ways in the classroom (Hyland, 2007; Peregoy & Boyle, 2016). To think deeper about fostering a culture of writing in mathematics through creating a communicative need, look at the statement written by Lourdes, a fifth-grade multilingual learner in Ms. Martínez's class, shown in Figure 10.1. In this statement, Lourdes describes what she perceives as the need for writing in mathematics.

Note that we transcribed the student work to facilitate its reading while trying to maintain the original words. You will find a transcript following the image of each piece of student work.

**Figure 10.1**    Lourdes's Writing Sample

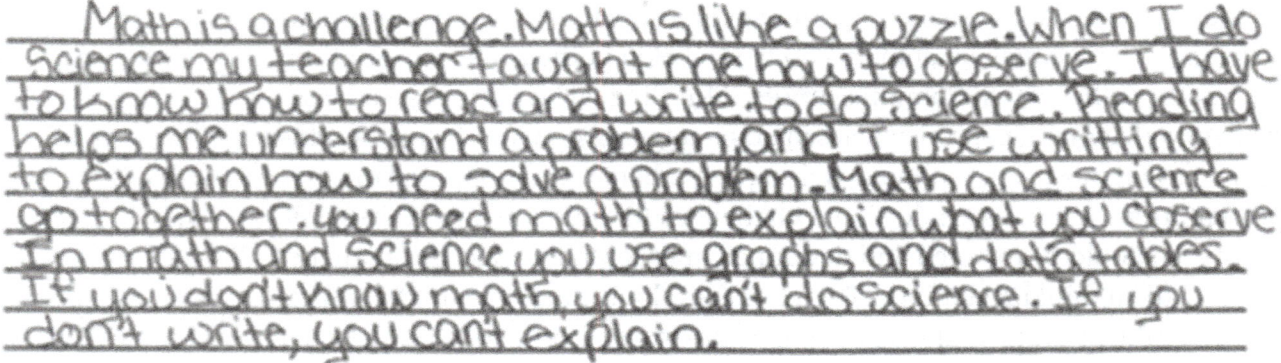

Math is a challenge. Math is like a puzzle. When I do science my teacher taught me how to observe. I have to know how to read and write to do science. Reading helps me understand a problem, and I use writing to explain how to solve a problem. Math and science go together. You need math to explain what you observe. In math and science you use graphs and data tables. If you don't know math, you can't do science. If you don't write, you can't explain.

Try It! 10.2 invites both you and your students to consider the need for writing in mathematics.

### Try It! 10.2

Ask your students why writing is needed in mathematics.

- What do you anticipate that they will say?

*You should establish and maintain high standards for all students, including multilingual learners, and respect and sustain students' heritage languages and their incorporation into their writing.*

Writing is a practice that must be cultivated. As such, writing must be done regularly and with purpose in a classroom environment that allows for diversity in writing. With that purpose in mind, you should establish and maintain high standards for all students, including multilingual learners, and respect and sustain students' heritage languages (García & Lin, 2017) and their incorporation into their writing. In the first days of the school year, Ms. Martínez began to establish her expectations for writing in her classroom:

> 66 *I want to share a couple of them [student writing samples] with you. I will not tell you who wrote them because I did not ask permission to read them . . . so I cannot identify the authors. But this is the kind of thing that I want you to think about all the time. We have to share things. That's how we learn. That's how we show people that we care for them.* 99

Notice how she connected sharing writing with thinking, learning, and caring for peers. From the very start of the school year, Ms. Martínez establishes that students will share their writings with each other for the purpose of becoming better writers. Further, it will be done in a respectful manner.

Multilingual learners need scaffolding as they begin to write in mathematics. In the first few weeks of school, Ms. Martínez's students began to learn about right triangles. They learned that a right triangle is composed of three sides (two sides that intersect to form the right angle are called the legs, and the longer remaining side is called the hypotenuse). After they learned how to find the area of a right triangle, Ms. Martínez posed more challenging problems. For instance, in one lesson she posed the problem, "Given the area of a right triangle and the length of one of the legs, what is the length of the other leg?" In this lesson, Ms. Martínez asked her students to write their explanation to the problem (of how to find the missing leg of a right triangle) for one of her friends. In Transcript 10.1 (adapted from Chval & Khisty, 2009), you can see how Ms. Martínez introduces the idea of her friend and her reason for showing the students' writing to her.

**Transcript 10.1**

| | |
|---|---|
| Ms. Martínez: | I want you to spend time writing about what we're doing. |
| Students: | Oh! |
| Ms. Martínez: | Oh, yes, that's hard, yes! I want to take this home, and I want to show it to one of my friends who doesn't know how to find the area of a right triangle, let alone a missing leg. |
| Juliana: | Is she an adult? |
| Ms. Martínez: | She's an adult. But she's a dancer, and dancers don't spend a whole lot of time writing about right triangles. But I want her to know how to do it, and I want her to read your papers, and if she understands what you wrote, then you've done a good job explaining. |

In this exchange, Ms. Martínez establishes a clear purpose for the writing assignment and why the students need to communicate effectively.

# ESTABLISHING A CULTURE OF WRITING

To develop a culture of writing, Ms. Martínez used a variety of approaches that we had not observed in other mathematics classrooms. She provided time for students to write every day. She made writing a public process by facilitating discussions before students wrote and then asking students to share their writing. Ms. Martínez created meaningful writing assignments that genuinely asked students to reflect about what they were learning. Rather than introducing many different writing assignments, she required students to write as many as six drafts of the same writing assignment in order to meet her expectations for a final product. As Ms. Martínez established a culture of writing, she immersed her students in an environment filled with words and learning interactions.

Ms. Martínez's teaching practices demonstrated that she valued writing in mathematics contexts. She modeled good writing as she wrote sentences on the chalkboard and as she wrote comments on student assignments; she explicitly and repeatedly discussed the importance of writing with her students; and she devoted time to it in her mathematics classes, in her preparation of lessons, and in the grading of student writing. Multilingual learners in Ms. Martínez's classroom heard the words in the context of solving mathematics problems and saw the words on the board and on written feedback. Through these practices, her students began to convey similar values as illustrated by Lourdes's statement in Figure 10.1.

## Providing Feedback on Students' Writing

Ms. Martínez understood the value of providing students written feedback on their mathematical writing and the benefits of revising written work. Yet, Ms. Martínez was strategic in the kinds of feedback she gave and focused her attention on specific aspects of their writing as students revised their written work. In Figures 10.2, 10.3, and 10.4, you can see student work in different stages of the revision process. Each of these examples is in response to the problem posed in Transcript 10.1 (i.e., writing an explanation to Ms. Martínez's friend on how to find the missing leg of a right triangle). As you examine each student's written work, reflect on the nature of Ms. Martínez's feedback.

In Figure 10.2, examine Javier's second draft of his explanation to Ms. Martínez's friend. On Javier's draft (from Chval & Khisty, 2009), you can see Ms. Martínez's questions to Javier in the cartoon bubbles. In the case of the first bubble, Javier responds by writing "triangle" after her question.

**Figure 10.2** Javier's Second Draft

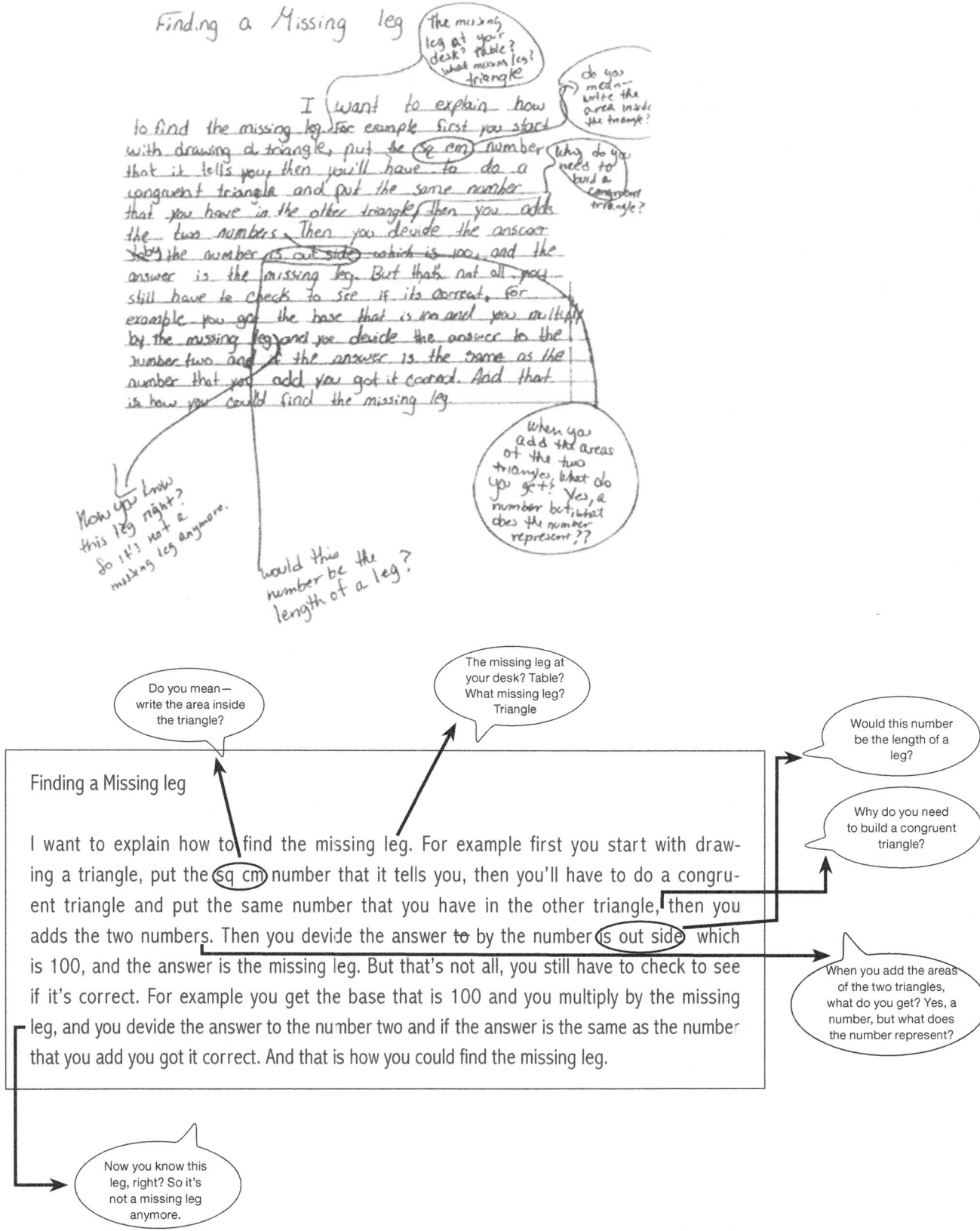

*Source:* Chval, K. B., & Khisty, L. L. (2009). Latino students, writing, and mathematics: A case study of successful teaching and learning. In R. Barwell (Ed.), *Multilingualism in Mathematics Classrooms: Global Perspectives*. Clevedon, UK: Multilingual Matters. Used with permission.

Look at Matthew's third draft in Figure 10.3 (from Chval & Khisty, 2009), which did not improve after Ms. Martínez's feedback on his second draft. In this round of feedback, Ms. Martínez is more direct in her expectations for the next draft.

**Figure 10.3**    Matthew's Third Draft

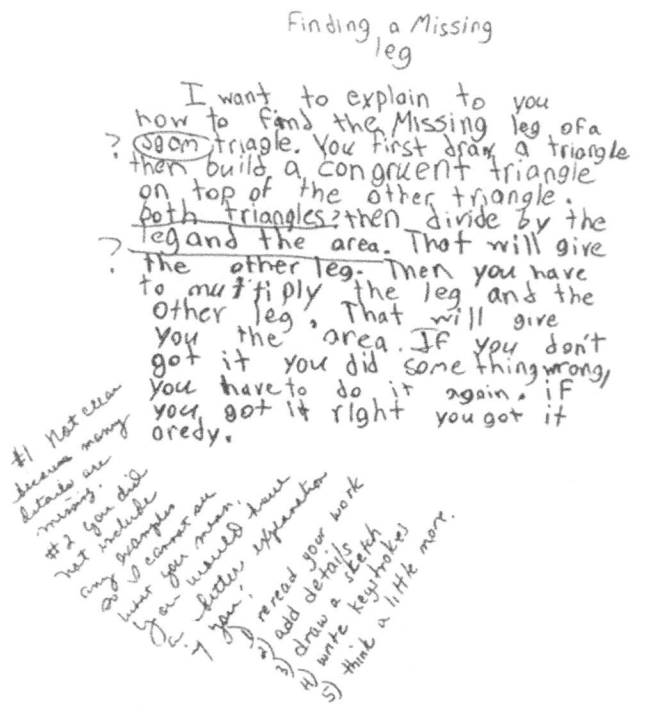

Finding a Missing leg

I want to explain to you how to find a missing leg of a sg cm? triangle. You first draw a triangle then build a congruent triangle on top of the other triangle. both triangles? then divide by the leg and the area?. That will give the other leg. Then you have to multiply the leg and the other leg. That will give you the area. If you don't got it you did something wrong, you have to do it again. If you got it right you got it aredy.

*Ms. Martínez's Feedback*

*#1 not clear because many details are missing*

*#2 you did not include any examples so I cannot see what you mean. You would have a better explanation if you:*

*1) reread your work*
*2) add details*
*3) draw a sketch*
*4) write keystrokes*
*5) think a little more.*

*Source:* Chval, K. B., & Khisty, L. L. (2009). Latino students, writing, and mathematics: A case study of successful teaching and learning. In R. Barwell (Ed.), *Multilingualism in Mathematics Classrooms: Global Perspectives*. Clevedon, UK: Multilingual Matters. Used with permission.

Alejandro's fourth draft is in Figure 10.4. In this draft, Ms. Martínez shifts the kind of feedback she provides to students as well as giving them praise on their efforts.

**Figure 10.4**   Alejandro's Fourth Draft

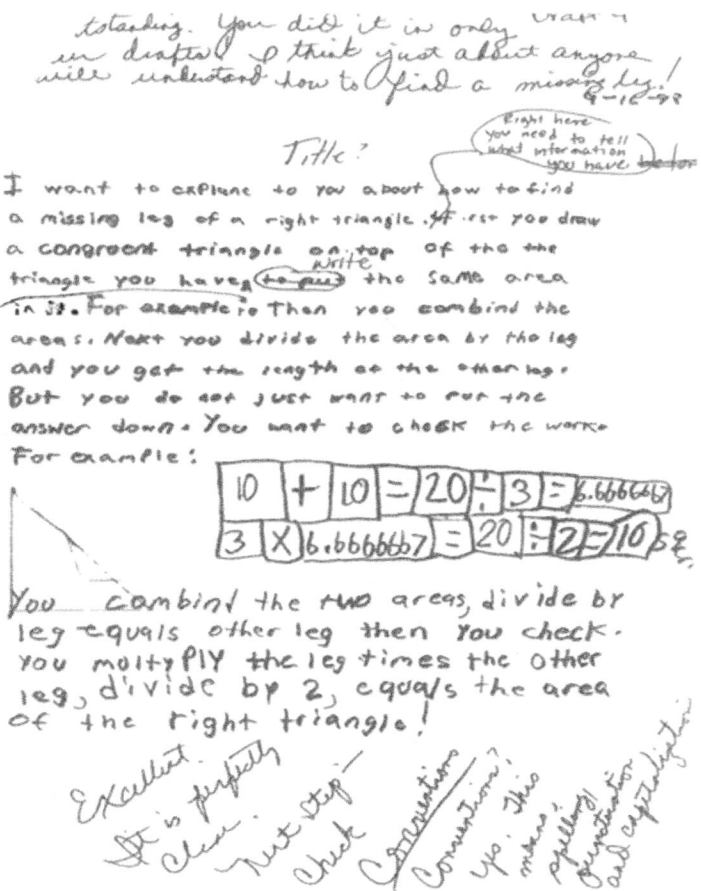

Ms. Martínez's Feedback: Outstanding. You did it in only four drafts! I think that just about anyone will understand how to find a missing leg!

Title?

> Right here you need to tell what information you have.

I want to explane to you about how to find a missing leg of a right triangle. First, you draw a congruent triangle on top of the the triangle you have ~~to put~~ write the same area in it. For example. Then you combind the areas. Next you divide the area by the leg and you get the length of the other leg. But you do not just want to put the answer down. You want to check the work. For example: [Keystrokes drawing]
You combind the two areas, divide by leg equals other leg then you check. You multiply the leg times the other leg, divide by 2, equals the area of the right triangle!

*Ms. Martínez's Feedback*

*Excellent! It's perfectly clear. Next step, check conventions! Conventions? Yes, this means spelling, punctuation and capitalization.*

## STOP AND THINK

Stop and think about Ms. Martínez's feedback in Figures 10.2, 10.3, and 10.4.

- How would you describe Ms. Martínez's feedback?

- How does her feedback differ based on what the student produced?

The writing process includes revision and feedback. Notice that Ms. Martínez used precise language, gave specific feedback, and focused attention on her students' strengths. In this way, she focused on building students' competencies in writing from the start. When providing feedback on students' writing, the main rule is to "first do no harm" (Jago, 2014). This means that you focus on the students' assets and ways of growing instead of looking only for errors. The writing process in the mathematics classroom should focus not on the product, but rather on the possibilities of growing students as mathematicians and writers.

*When providing feedback on students' writing, the main rule is to "first do no harm."*

As you examined Ms. Martínez's feedback, you likely noticed that she used sophisticated written language with her multilingual fifth graders, such as this:

- ▶ *Verify* your results.

- ▶ *Combine* the areas.

- ▶ What does that number *represent*?

- ▶ Your work is *deteriorating*.

- ▶ Use your *power of observation*.

- ▶ *Clarify* this example.

- ▶ Next step, check *conventions*.

- ▶ This *procedure* does not help me find the missing leg.

*Source:* Chval and Khisty (2009).

*If precise mathematical language is part of students' daily discourse, then mathematical terms become part of their language repertoire and discourse—both multilingual learners and monolingual learners alike.*

Some teachers have the tendency to reduce the curriculum's level of complexity to multilingual learners' English proficiency (González, Moll, & Amanti, 2005). Notice that Ms. Martínez does not; as she stated to her students, "If I don't use the words, then the students won't hear them." If precise mathematical language is part of students' daily discourse, then mathematical terms become part of their language repertoire and discourse—both multilingual learners and monolingual learners alike.

You probably also noted that Ms. Martínez wrote questions on student assignments leading students to be more explicit about their ideas. This practice facilitated conversations with her students via writing, as noted when Javier replied to her question with the word *triangle*. To grade students' final written products, Ms. Martínez uses a rubric that includes five components: Focus, Support, Organization, Conventions, and Overall. (Sample rubrics are available at https://cwp.missouri.edu/instructor/wiassignments/.)

## STOP AND THINK

Stop and think about the different components in Ms. Martínez's grading rubric.

- Why are these components important for multilingual learners and their mathematical writing process?

## ANALYZING STUDENTS' MATHEMATICAL WRITING

Analyzing students' mathematical writing provides so many opportunities to understand student thinking and consider ways to enhance our teaching practices and design our mathematical tasks. In Figures 10.5, 10.6, and 10.7, you will find three different mathematical writing tasks and respective student work. Examine these samples from Ms. Martínez's classroom, paying attention to what information you can glean from the students' written work.

**Figure 10.5** Student A

Is it possible to have a right triangle that has sides 2 cm, 4 cm, and 6 cm? Why or why not?

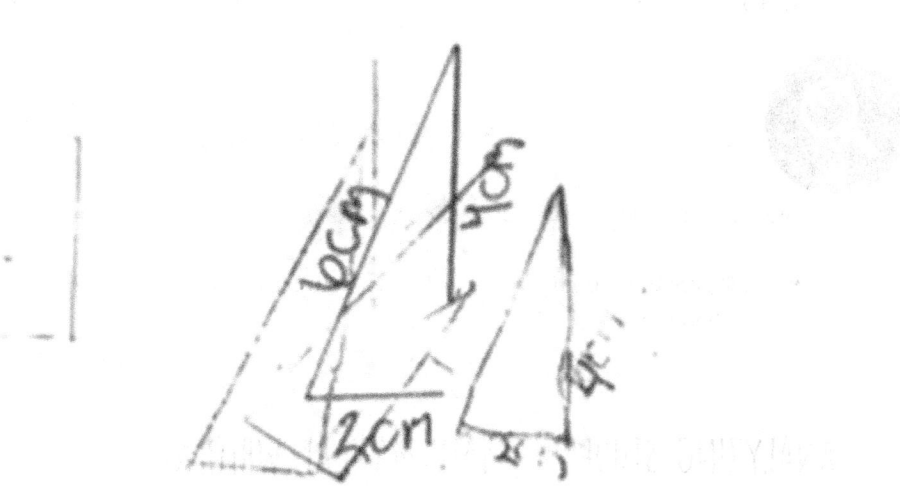

Mis. Chav

I know your thinking
it is that the triagle it
dose not. Look like a
Rright triage. Because 4 and 6
it could no be togeter beacuas
4 is it to smalle for it.

**Figure 10.6**   Student B

Explain how to find the missing leg of a right triangle given the area of the triangle and the length of one leg.

I want to shoul you how to find the messing leg wen you have a right. triangle leg messing. First you braw a congrue triangle. Thene you add areas of both triangles. Then it wil give you the whole rectangle area, Then You divide the leg that they give you, and they will give you the other leg. If you what to make shour that its right you moiltiply the two legs. It will give you the whole rectangle. But you don't want the whole rectangle so you Divided by 2, and it wil give you the right triangle, and these how you do It-

---

Finding a MESinj Leg

I want to shoul you how to find the messing leg wen you have a right triangle leg messing. First you braw a congrue triangle. Thene you add areas of both triangles. Then it wil give you the whole rectangle area. Then you divide the leg that they give you, and they will give you the other leg. If you want to make shour that its right you multiply the two legs. It will give you the whole rectangle. But you don't want the whole rectangle so you Divided by 2, and it wil give you the right triangle, and these how you do it.

---

*Source:* Chval, K. B., & Khisty, L. L. (2009). Latino students, writing, and mathematics: A case study of successful teaching and learning. In R. Barwell (Ed.), *Multilingualism in Mathematics Classrooms: Global Perspectives.* Clevedon, UK: Multilingual Matters.

**Figure 10.7**  Student C

If the area of a $\frac{3}{4}$ circle is 100 cm², find the perimeter of the $\frac{3}{4}$ circle.

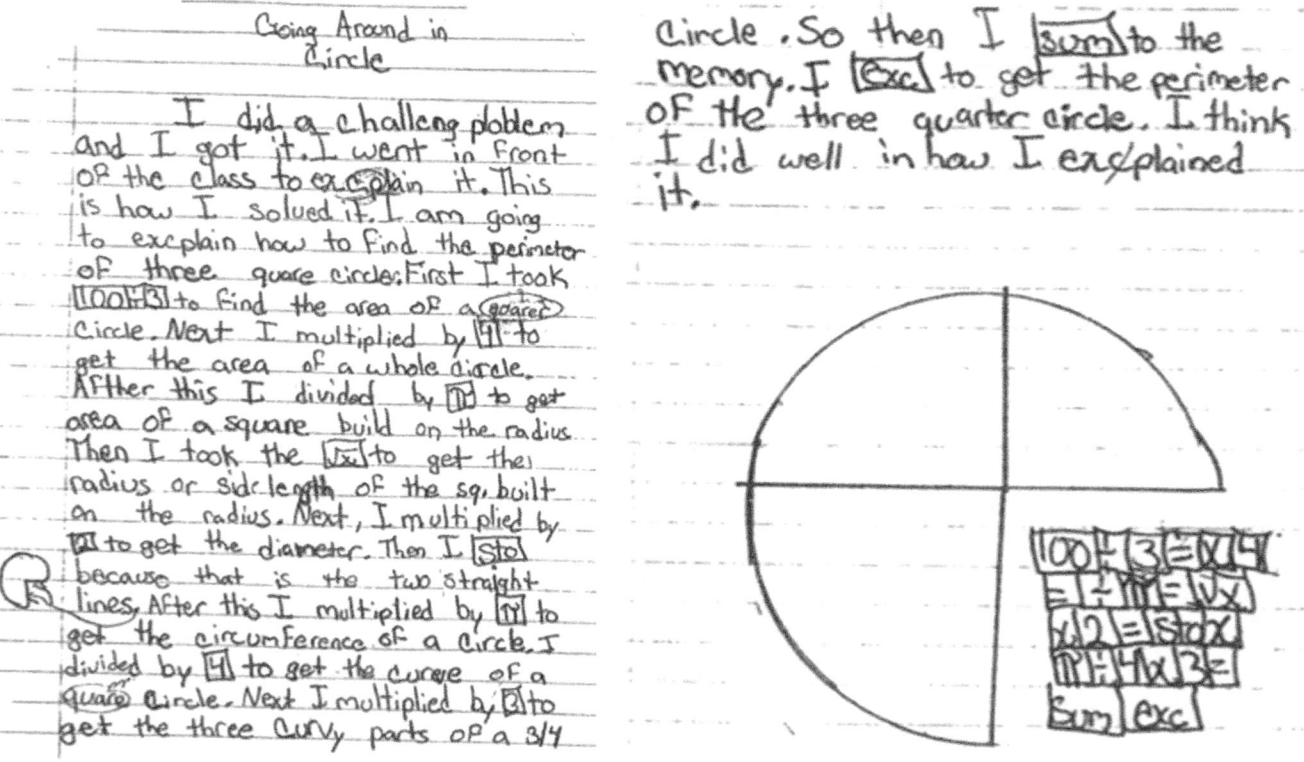

---

Going Around in Circle

I did a challeng ploblem and I got it. I went in Front of the class to excplain it. This is how I solved it. I am going to excplain how to find the perimeter of three quare circles. First I took **100 ÷ 3** to find the area of a quarer circle. Next I multiplied by 4 to get the area of a whole circle. Afther this I divided by π to get area of a square build on the radius. Then I took the $\sqrt{x}$ to get the radius or side length of the sq. built on the radius. Next, I multiplied by **2** to get the diameter. Then I **sto** because that is the two straight lines. After this I multiplied by **π** to get the circumference of a circle. I divided by **4** to get the curve of a quare circle. Next I multiplied by 3 to get the three curvy parts of a 3/4 circle. So then I **sum** to the memory. I **exc** to get the perimeter of the three quarter circle. I think I did well in how I excplained it.

$$100 ÷ 3 = × 4$$
$$= ÷ π = \sqrt{\phantom{x}}$$
$$× 2 = sto ×$$
$$π ÷ 4 × 3 =$$
**sum exc**

---

*Source:* Chval, K. B., & Khisty, L. L. (2009). Latino students, writing, and mathematics: A case study of successful teaching and learning. In R. Barwell (Ed.), *Multilingualism in Mathematics Classrooms: Global Perspectives.* Clevedon, UK: Multilingual Matters. Used with permission.

## STOP AND THINK

Stop and think about the writing samples in Figures 10.5, 10.6, and 10.7.

- How do the three writing samples differ?

- Were these samples written by the same student or by different students? What evidence did you use?

You likely noticed that Student C embedded some calculator keystrokes within the narrative. Ms. Martínez required students to list the calculator keystrokes they used to solve challenging problems; however, the students extended this practice and began to include keystrokes in their narratives over the course of the year.

You guessed it. All three samples, A, B, and C, were written by Violetta, a fifth-grade multilingual Latina. Sample B was written during the first week of school. Sample A was written during the third week of school. Sample C was written in April, toward the end of the fifth-grade year. As demonstrated by Sample C, Violetta advanced as a writer over the course of the school year.

Ms. Martínez created a need for mathematical writing and scaffolded the writing process. Her multilingual learners had opportunities to critically reflect on their writing, to understand the purpose for writing, and to foster the development of their writing community. In addition, you can see how the samples represent different kinds of writing, or genres, that correspond to different purposes. In Ms. Martínez's classroom, instruction that facilitates writing for multilingual learners has clear benefits for her students:

- It empowers students to think like mathematicians and to think of themselves as mathematicians.

- It nurtures students' ability to critically reflect on their writing and improve their drafts over time.

- It provides a foundation for students to understand the importance and purpose of clear and effective writing.

- It confirms the idea that students can become expert writers in a variety of genres.

In Chapter 11, you will analyze different mathematical writing genres and explore teaching activities to support mathematical writing genres for multilingual learners.

## STRATEGIES FOR FOSTERING A CULTURE OF WRITING IN YOUR MATHEMATICS CLASSROOM

As you work to foster a culture of writing in your mathematics classroom, consider the following strategies and ideas:

> ▶ A culture of writing in the classroom allows students to discover "what they know about their content focus, their language, themselves, and their ability to communicate all that to a variety of audiences" (Knipper & Duggan, 2006, pp. 462–463). You can support the shared construction of written texts in the mathematics class. You can also contribute to the construction of mathematical texts with your students.

> ▶ Express your expectations about writing in mathematics from the beginning of the school year so that multilingual learners know how to approach writing in your class.

> ▶ Focus on the expression of thinking processes and ideas before emphasizing form or conventions (Hyland, 2019).

> ▶ Build confidence and trust in your multilingual learners through a culture of writing (Hyland, 2007).

> ▶ As students develop as writers, highlight the importance of context, purpose, and need in their writing activities (Brisk, 2011; Hyland, 2007, 2019). This encourages students to consider the kind of language they use (e.g., formality, terminology) and to focus on the content that will help the reader achieve the goal (e.g., understand the mathematical idea sufficiently, grasp an activity parents could do; Gebhard, Chen, & Britton, 2014; Hyland, 2007). If students have the possibility to submit their writing to a real audience (as opposed to just you), this is even better (Graham, McKeown, Kiuhara, & Harris, 2012).

> ▶ Although writing is usually an individual activity, writers grow when they share their writing and receive feedback from others. However, you cannot assume students will know how to provide constructive feedback to peers in a respectful manner. It may be beneficial to explicitly discuss what is productive and unproductive with students. When teachers establish a culture that values writing in the mathematics classroom, their students will grow in both mathematics and writing. As we saw in Chapter 9, you are a model of English language and the language of mathematics for the multilingual learners in your classroom (Chval & Khisty, 2001). What students read and hear from you is what they will eventually acquire as their academic language and will subsequently use for written communication.

> ▶ Multilingual learners benefit from prewriting activities, such as "brainstorming, discussing, and being shown a model," prior to writing (Davison & Pearce, 1990, p. 19; Zheng & Dai, 2012).

▶ Informal writing is essential for students to develop their mathematical ideas. Throughout this process, students should focus on their mathematical thinking, rather than a final product (Wilde, 1991).

▶ When providing feedback to multilingual learners, you should use an asset-based approach with a primary focus on what the student "can [do] and has done" (Jago, 2014, p. 10) rather than focusing on errors. It is important you provide students with supportive and critical feedback in an encouraging way.

▶ Respect and sustain multilingual learners' heritage languages by allowing students to use multiple languages in their texts; this is called translanguaging. That is, multilingual learners can use their multiple language repertoires and expressions, in English and in their heritage or first languages, to create and negotiate meaning with the text (García & Lin, 2017; Velasco & García, 2014). You can use this strategy for students with any language background.

▶ Include activities where multilingual learners can use cultural information about their families and communities and their knowledge of the world in their mathematical writing. This provides students an opportunity to think about their knowledge in other languages and the contexts and audiences that interest them (García & Lin, 2017).

## REFLECTING ON MATHEMATICAL WRITING IN YOUR PRACTICE

Consider how the culture of mathematical writing has been developed in your school. Then reflect on the questions in Try It! 10.3.

 **Try It! 10.3**

- How is writing used to record important mathematical ideas that are under investigation and discussed?

- How are teachers focusing students on mathematical writing?

- How are teachers focusing students on analyzing their own and others' writing?

- What spaces are made in the classroom (physical and metaphorical) for mathematical writing?

### Reflect

- What next steps will you take to foster a culture of writing in your mathematics classroom?

# CHAPTER 11
## DEVELOP WRITING IN MATHEMATICS FOR MULTILINGUAL LEARNERS

### Key Concepts

In this chapter, you will

✓ learn about different genres of mathematical writing.

✓ learn strategies for engaging multilingual learners in writing different genres.

In Chapter 10, you learned about the importance of developing a culture of writing in the mathematics classroom. In addition, you identified strategies to foster a culture of writing. In this chapter, we shift our attention to strategies you can use to engage multilingual learners in a range of mathematical writing activities. As we interacted with teachers, they noticed patterns in the types of writing they required from their students. For example, they asked students to "show your work" or "explain the different ways that you could approach this problem." However, they did not necessarily include writing tasks where their multilingual learners justified their thinking, or critiqued the reasoning of others. In other words, they recognized that they included some writing genres (i.e., different forms of texts), but not others, in their lessons.

## REFLECTING ON YOUR PRACTICE

There are a range of things students are asked to write about in mathematics. For instance, they might be asked to explain a strategy, describe their thinking, or provide a mathematical justification. Each of these is important because it supports students' mathematical thinking.

### STOP AND THINK

Stop and think about mathematical writing in your classroom.

- What do you ask multilingual learners to write about in mathematics?

## MATHEMATICAL GENRES OF WRITING

Throughout your typical day, you may encounter different types of writing—for example, a shopping list; directions; assignments; and emails or letters to parents, students, colleagues, and administrators. Look at the following list and think about how these types of writing differ.

- Writing an editorial for the school journal
- Writing a Supreme Court ruling
- Writing a letter to a parent
- Writing a memo to terminate an employee
- Writing a mathematical proof
- Writing a mathematical word problem

You may have noted that each piece of writing has a different

- point of view (e.g., first person);
- audience (e.g., lawyers, mathematicians, students);

> purpose (e.g., to convince the reader through an argument);

> vocabulary/register (e.g., audience-appropriate vocabulary and tone);

> length (e.g., a few sentences vs. multiple pages);

> structure (e.g., argument, description, narrative); and

> density (i.e., the amount of cognitive load required to read and interpret).

Each of these writing forms constitutes a different *genre* because it has a determined structure; certain kind of language (e.g., legal language vs. specialized mathematical language); level of formality; purpose, audience, content, organization, and style. A writing genre is a category under which you can identify texts with certain shared characteristics, conventions, and structures. Writing genres are popular in literature, such as comedy, drama, and satire. However, writing genres are broader, and you can identify them in all forms of writing.

*A writing genre is a category under which you can identify texts with certain shared characteristics, conventions, and structures.*

## WHAT THE RESEARCH SAYS ABOUT MATHEMATICAL GENRES

As you saw in Chapter 10, developing the ability to communicate mathematically has great benefits for every learner, particularly multilingual learners. Many cognitive benefits have been attributed to writing in content areas for multilingual learners (Hirvela, 2011), including enhanced communication and understanding of mathematical ideas (Freeman, Higgins, & Horney, 2016). Students need deliberate and explicit experiences to learn the different competencies involved in mathematical communication. One way to communicate mathematically is through writing, which is a complex ability. An effective writer needs to know the appropriate mathematical genre and specialized language of mathematics (Accurso, Gebhard, & Purington, 2017; Chval & Khisty, 2009; Schleppegrell, 2004).

*Developing the ability to communicate mathematically has great benefits for every learner, particularly multilingual learners.*

Like other content areas, mathematics has writing genres. Marks and Mousley (1990) describe them as follows:

" Mathematicians use language to make meanings and to share their understandings. Discourse in mathematics involves oral and written communication, gesture, the use of drawings and diagrams and other language forms . . . . In mathematics, as in other expressions of language, there are accepted structural forms, which are used to make meaning. The social contexts in which texts are generated are of fundamental importance. These occasions become predictable, leading to the generation of conventionalized forms of texts, or *genres*. (pp. 118–119) "

Thus, mathematics has a set of conventional writing genres like other types of writing. For example, in mathematics, teachers often ask students to explain a mathematical procedure or justify a mathematical strategy. Each of these different kinds of prompts requires a different mathematical genre and is common in curricula and on standardized assessments. Let's delve deeper into the different genres in mathematical writing by analyzing samples of students' writing.

## ANALYZING STUDENTS' MATHEMATICAL WRITING

The following five writing assignments, shown in Figures 11.1, 11.2, 11.3, 11.4, and 11.5, were completed by multilingual learners from Ms. Martínez's fifth-grade class. Remember that this class was 100% multilingual learners. As you examine the individual writing tasks and student writing samples, identify how they are similar and different. As in Chapter 10, we transcribed the student work to facilitate its reading while trying to maintain the original words. You will find a transcript after the image of each sample of student work.

**Figure 11.1**  Sample 1

Is it possible to draw a right triangle that has sides 2, 4, and 6 cm? Why or why not?

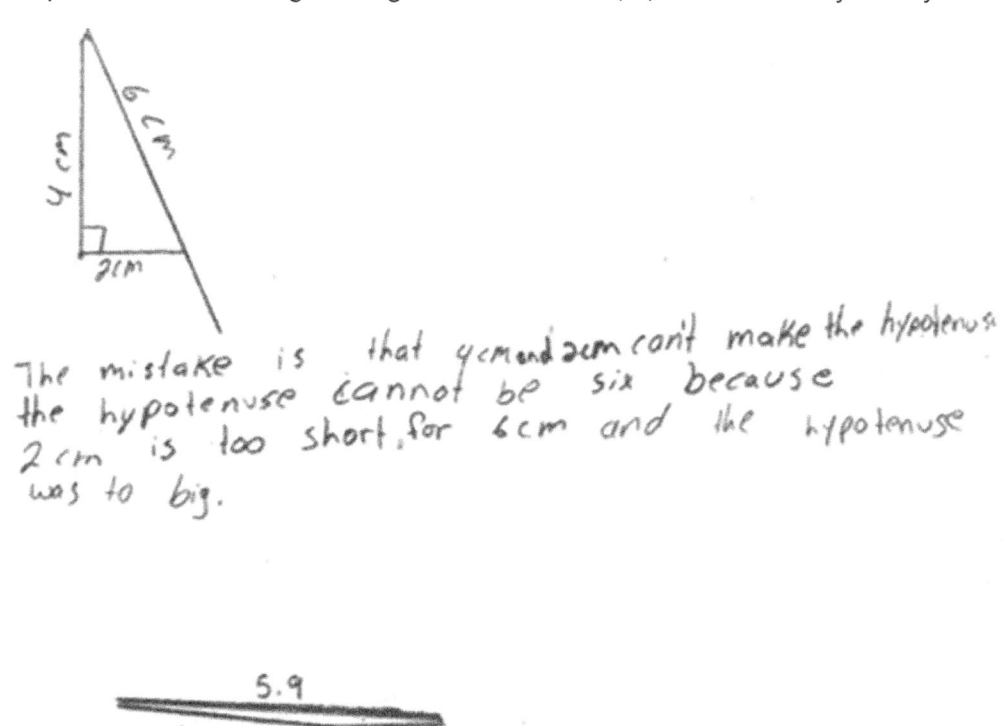

The mistake is that 4 cm and 2 cm can't make the hypotenuse

the hypotenuse cannot be six because

2 cm is too short for 6 cm and the hypotenuse

was to big.

**Figure 11.2**   Sample 2

What is area? How do you calculate area for a right triangle?

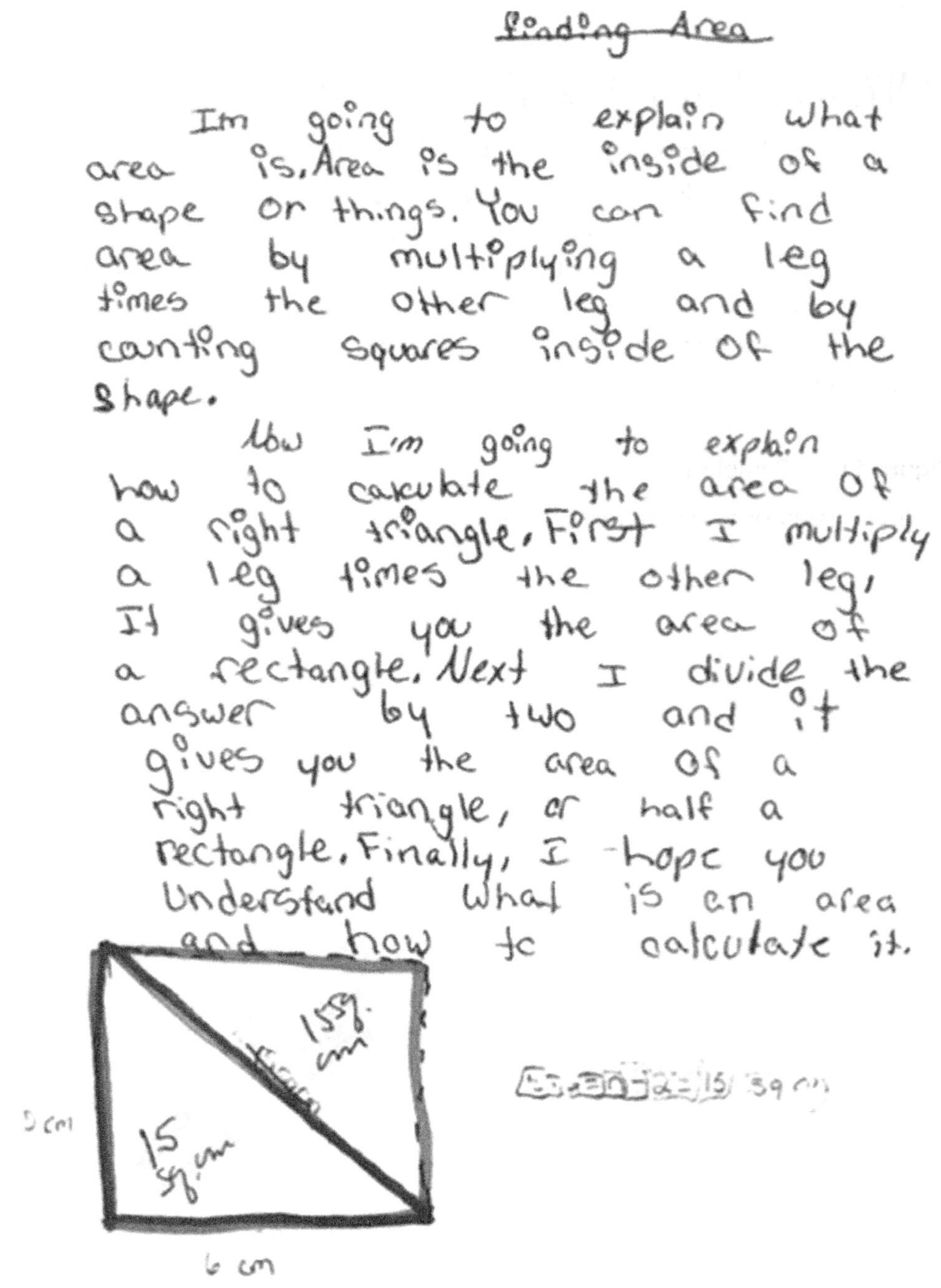

Finding Area

I'm going to explain what area is. Area is the inside of a shape or things. You can find area by multiplying a leg times the other leg and by counting squares inside of the shape.

Now I'm going to explain how to calculate the area of a right triangle. First I multiply a leg times the other leg, It gives you the area of a rectangle. Next I divide the answer by two and it gives you the area of a right triangle, or half a rectangle. Finally, I hope you understand what is an area and how to calculate it.

---

Finding Area

I'm going to explain what area is. Area is the inside of a shape or things. You can find area by multiplying a leg times the other leg and by counting squares inside of the shape.

Now I'm going to explain how to calculate the area of a right triangle. First I multiply a leg times the other leg, It gives you the area of a rectangle. Next I divide the answer by two and it gives you the area of a right triangle, or half a rectangle. Finally, I hope you understand what is an area and how to calculate it.

**Figure 11.3** Sample 3

Explain the differences between categorical and numerical variables.

Caterigorical Numerical

Categorical means that what day you are born or what is your favorite hobby. Numerical is that the answer should have numbers and the Categorical should not have numbers. Like someone is telling you what is your birthay day and I say wednesday, and I think about is it Categorical or Numerical !aha a! is Categorical, or someone say's how many years old is your nephew( and I say 1 year old, and I think about it and I say is it Categorical or Numerical !ahaa! is numerical. That is the explanation of Numerical and Categorical.

Caterigorical Numerical

Categorical means that what day you are born or what is your favorite hobby. Numerical is that the answer should have numbers and the categorical should not have numbers. Like someone is telling you what is your birthay day and I say Wednesday, and I think about is it Categorical or Numerical !ahaa! is Categorical, or someone say's how many years old is your nephew and I say 1 year old, and I think about it, and I say is it Categorical or Numerical !ahaa! is numerical. That is the explanation of Numerical and Categorical.

**Figure 11.4** Sample 4

Explain how you find the surface area of a three-dimensional object.

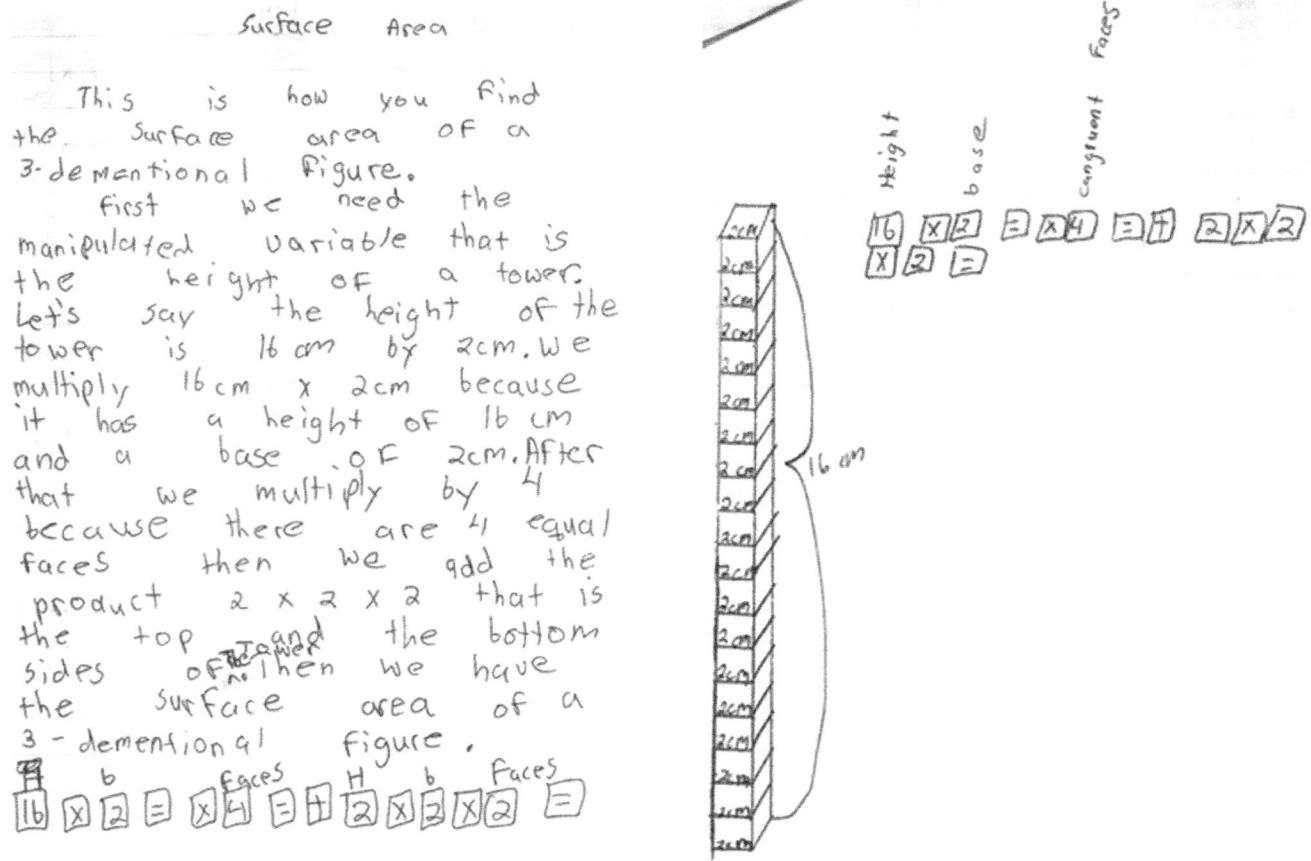

Surface Area

This is how you find the surface area of a 3-dementional figure.

First we need the manipulated variable that is the height of a tower. Let's say the height of the tower is 16 cm by 2 cm. We multiply 16 cm × 2 cm because it has a height of 16 cm and a base of 2 cm. After that we multiply by 4 because there are 4 equal faces then we add the product 2 × 2 × 2 that is the top and the bottom sides of the Tower. Then we have the surface area of a 3-dementional figure.

**Figure 11.5**   Sample 5

Is the area of the three squares more than, less than, or equal to the area of the circle? Why?

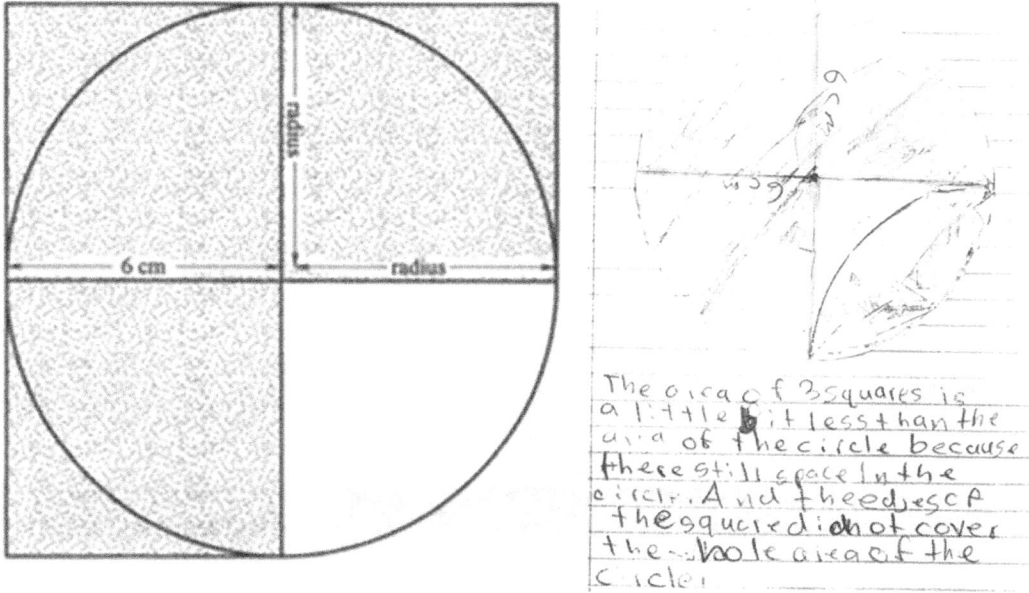

*Source:* Page, D. A., Wagreich, P., & Chval, K. (1993). *Maneuvers with circles.* Palo Alto, CA: Dale Seymour. Reprinted with permission.

> The area of 3 squares is a little bit less than the area of the circle because there still space In the circle. And the edjes of the square didnot cover the whole area of the circle.

## STOP AND THINK

Stop and think about the writing prompts that Ms. Martínez assigned.

- How do the writing prompts differ?

- Based on the student writing samples, how did the use of different prompts facilitate students' learning through mathematical writing?

- What types of writing prompts do you use in your mathematics classroom?

Ms. Martínez used a range of writing prompts to produce the different types of writing. Some prompts students had experienced before (e.g., prompts 3 and 4), whereas others they had not (e.g., prompt 1). In addition, the prompts were provided at different times of the school year. For instance, prompt 3 was given to students in August, at the start of the school year, to establish writing expectations. In contrast, prompt 5 was given in January for students to build meaning for *pi*.

As you reviewed the writing samples, you may have noticed differences in the purpose, targeted audience, teacher expectations, and rhetorical strategies used (e.g., different uses of voice and tone; expressions such as "first," "second," "finally," "let's say that," and "I hope you understand"). In addition, you may have noticed how the multilingual learners in Ms. Martínez's class refined their writing proficiencies and the range of things they can do and say over the course of the year. As discussed in Chapter 10, it is important to acknowledge that Ms. Martínez emphasized mathematical writing on a regular basis and shared students' writing during class to support her students' learning.

## IDENTIFYING MATHEMATICAL WRITING GENRES

In Try It! 11.1 (adapted from Marks & Mousley, 1990), you will find a list of mathematical writing prompts. For each row, write in the right-hand column the characteristics that describe each pair of writing prompts. See the first row as an example.

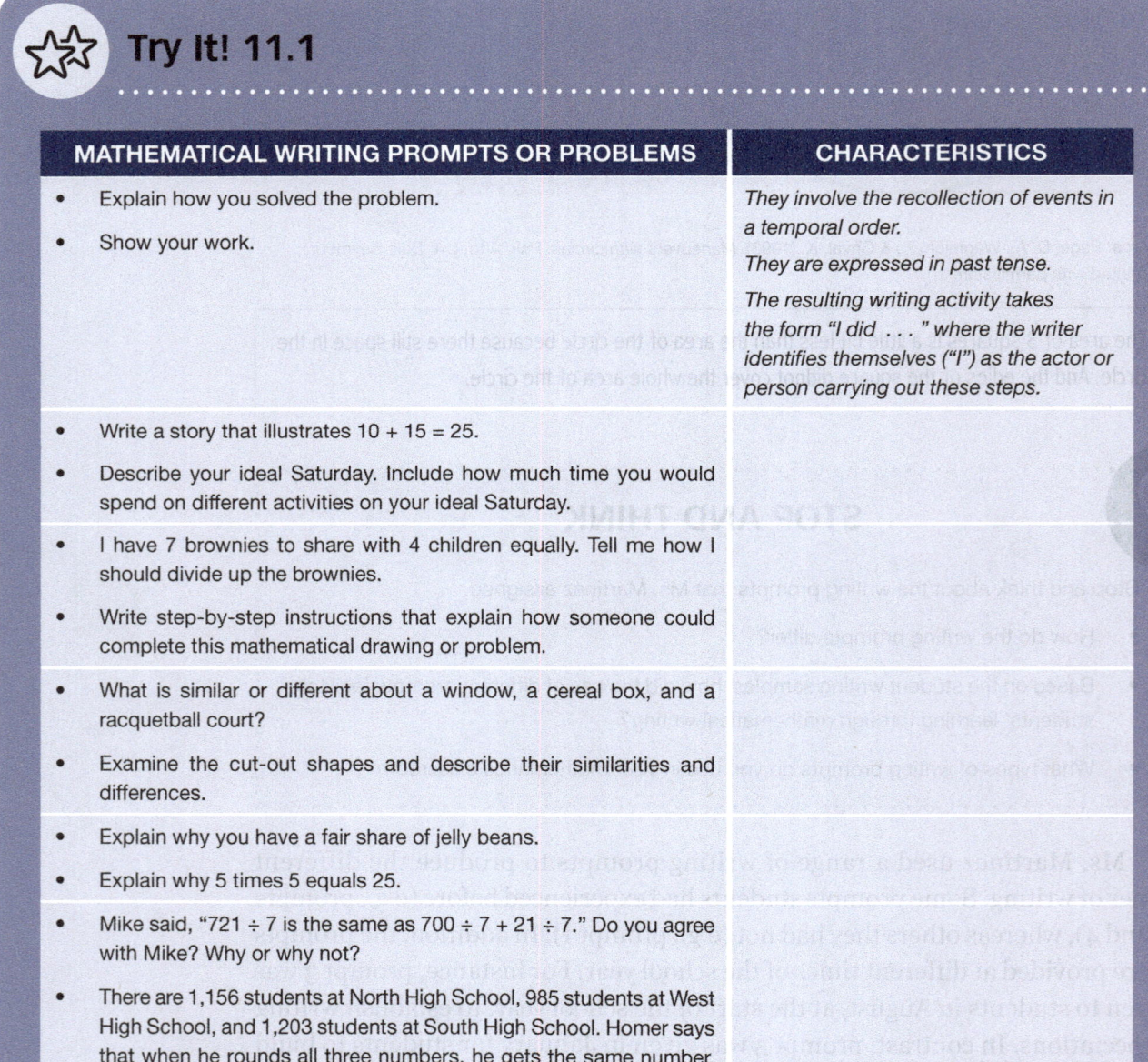

### Try It! 11.1

| MATHEMATICAL WRITING PROMPTS OR PROBLEMS | CHARACTERISTICS |
|---|---|
| • Explain how you solved the problem.<br>• Show your work. | *They involve the recollection of events in a temporal order.*<br>*They are expressed in past tense.*<br>*The resulting writing activity takes the form "I did . . . " where the writer identifies themselves ("I") as the actor or person carrying out these steps.* |
| • Write a story that illustrates 10 + 15 = 25.<br>• Describe your ideal Saturday. Include how much time you would spend on different activities on your ideal Saturday. | |
| • I have 7 brownies to share with 4 children equally. Tell me how I should divide up the brownies.<br>• Write step-by-step instructions that explain how someone could complete this mathematical drawing or problem. | |
| • What is similar or different about a window, a cereal box, and a racquetball court?<br>• Examine the cut-out shapes and describe their similarities and differences. | |
| • Explain why you have a fair share of jelly beans.<br>• Explain why 5 times 5 equals 25. | |
| • Mike said, "721 ÷ 7 is the same as 700 ÷ 7 + 21 ÷ 7." Do you agree with Mike? Why or why not?<br>• There are 1,156 students at North High School, 985 students at West High School, and 1,203 students at South High School. Homer says that when he rounds all three numbers, he gets the same number. Alex says that when he rounds all three numbers, only two are the same. Are Homer or Alex correct? Why or why not? | |

*Source:* Adapted from Marks and Mousley (1990).

Now, look back over your responses in the right-hand column of Try It! 11.1. How do each of the writing prompts differ in their characteristics?

In Figure 11.6 (adapted from Marks & Mousley, 1990), we detail the characteristics and associated mathematical genres for each of the prompts from Try It! 11.1. How do your responses in Try It! 11.1 compare to ours? After comparing your responses to ours, highlight the genres in Figure 11.6 that you want to incorporate more in your mathematics instruction.

**Figure 11.6**  Mathematical Writing Genres

| MATHEMATICAL WRITING PROMPTS | GENRE CHARACTERISTICS | MATHEMATICAL GENRES |
|---|---|---|
| • Explain how you solved the problem.<br>• Show your work. | It involves the recollection of events in a temporal order.<br>It is expressed in past tense.<br>It takes the form "I did . . ." | Recount genre |
| • Write a story that illustrates 10 + 15 = 25.<br>• Describe your ideal Saturday. Include how much time you would spend on different activities. | It involves telling a story.<br>It takes the form of a narrative text.<br>It can include writing a story problem. | Storytelling genre |
| • I have 7 brownies to share with 4 children equally. Tell me how I should divide up the brownies.<br>• Write step-by-step instructions that explain how someone could complete this mathematical drawing or problem. | It is used to explain how to do something that has not been done yet.<br>It is expressed in imperative mood (e.g., "Take away . . ."). | Procedural genre |
| • What is similar or different about a window, a cereal box, and a racquetball court?<br>• Examine the cut-out shapes and describe their similarities and differences. | It uses the language of actual inquiry to describe an entire class of objects. It often involves understanding acquired through observation of the real world.<br>It requires attention to detail, sequencing, situation-specific description, comparison, and concise definition. | Report genre |
| • Explain why you have a fair share of jelly beans.<br>• Explain why 5 times 5 equals 25. | It involves explaining the reasoning behind a computation or a result. | Explanatory genre |
| • Mike said, "721 ÷ 7 is the same as 700 ÷ 7 + 21 ÷ 7." Do you agree with Mike? Why or why not?<br>• There are 1,156 students at North High School, 985 students at West High School, and 1,203 students at South High School. Homer says that when he rounds all three numbers, he gets the same number. Alex says that when he rounds all three numbers, only two are the same. Are Homer or Alex correct? Why or why not? | It requires the use of arguments to demonstrate why one answer is correct, or to assume a posture toward a problem, or to evaluate different options.<br>A claim is proven using logical reasoning and examples. | Argumentative genre |

*Source:* Adapted from Marks and Mousley (1990).

Because each mathematical writing genre focuses on developing different thinking skills and requires different thinking processes, you should aim to include a range of mathematical writing genres in your lessons with multilingual learners. As you design your classes, recognize that multilingual learners are in the process of learning the dynamics of the English language across contexts. Their writing will often display phonetic spelling and other developmental features of a second language writer. This does not mean that writing should be avoided since we learn to write by *writing*. Also, keep in mind that any language domain (i.e., writing, reading, listening, or speaking) that is neglected will not develop as it should.

### STOP AND THINK

Stop and think about the mathematical genres that you prioritize in the classroom.

- Are there specific genres you emphasize more than others?
- Are there any genres you avoid?

For Try It! 11.2, look back at one of the mathematical genres you highlighted in Figure 11.6.

### Try It! 11.2

Review the upcoming topics you will teach and create a writing prompt in the mathematical genre you selected from Figure 11.6. Remember, it's okay to start with a simple prompt and add complexity as you and your students advance. What is most important is that you begin to incorporate a range of mathematical writing genres and activities.

## STRATEGIES FOR DEVELOPING WRITING IN MATHEMATICS FOR MULTILINGUAL LEARNERS

As a teacher, there are a multitude of ways to incorporate into your classroom a range of mathematical writing activities that span genres. Here are some strategies you can use with multilingual learners in the classroom:

### Teaching Practices

▶ Prior to writing, students should discuss their ideas orally. By allowing students opportunities to hear and speak the language, they will begin the process of connecting the symbolic notation of language to its meanings (Chval, 2004; Rojas-Drummond et al., 2017).

▶ Multilingual learners benefit from writing activities that focus on the process more than on the product—that is, focusing on thinking, drafting, reviewing, and writing (Hyland, 2019). Through these writing activities, the teacher's role shifts from assessing a final product to one that provides support throughout the entire process (Peregoy & Boyle, 2016).

▶ Students who have experience writing in mathematics classrooms should be tasked with writing to "explore process problems—those that must be figured out through reasoning, trial and error, or insight rather than by using an algorithm" (Wilde, 1991, p. 41). When first introducing students to this type of writing, you should model through your own writing (Wilson & Devereux, 2014).

## Activities

▶ Ask your multilingual learners to write word problems (Kersaint, Thompson, & Petkova, 2014). This activity helps students learn the features and characteristics of word problems, allows students to bring their own backgrounds and interests into problem contexts, and fosters the integration of mathematics and language (Barwell, 2009). Writing word problems benefits not only multilingual learners but all your students because it involves high-level cognitive abilities and can help improve the students' skills to solve problems (Rothstein & Rothstein, 2007).

▶ Use writing as a diagnostic tool to evaluate student knowledge and attitudes toward mathematics. To assess students' attitudes toward mathematics at the start of the school year, ask students to write what mathematics means to them (Wilde, 1991). Based on student responses, you can gain insight into students' prior mathematical experiences and how they conceptualize the value of mathematics.

▶ Provide students with rubrics to assess their own and others' writing. By reviewing their own and others' writing against agreed-upon public criteria, students can see examples and offer critiques (Knipper & Duggan, 2006; Panadero & Jonsson, 2013).

▶ Ask students to keep a mathematical diary or journal. It provides a platform for students to represent, demonstrate, explain, and reflect on their thinking (Yang, 2005). Further, it provides a direct, private communication link between the teacher and the student. When you respond to student journals on a frequent basis, you establish a closer relationship with your multilingual students. Some examples of prompts are provided as follows (Dougherty, 1996):

> ▶ *Content prompts.* Ask students to describe or compare and contrast a concept or topic. You can also ask students to discuss the evolution of their mathematical ideas over time.

▶ *Process prompts.* Ask students to explain why they choose or prefer a way of solving a problem, to describe what their study habits are, or to reflect on their problem-solving processes.

▶ *Affective prompts.* Ask students to explain their feelings or attitudes and beliefs toward mathematics or a specific topic. You can also ask students to write about their view of themselves as mathematical doers.

▶ Provide students with a piece of student work. Ask them to analyze it. For example, the following task is from Lannin, Chval, and Jones (2013, p. 162):

---

Kara says that when you multiply two numbers, the answer is always bigger.

Do you agree with Kara?          Circle:               Yes               No

Explain your thinking.

---

## REFLECTING ON YOUR USE OF MATHEMATICAL GENRES IN YOUR PRACTICE

Consider how you can create writing opportunities or find writing prompts in your curriculum materials. Then reflect on the questions in Try It! 11.3.

### Try It! 11.3

Select one or two important topics or chapters in your curriculum. Look at the topics and chapters with the following questions in mind:

1.   What kinds of mathematical writing opportunities are available to students?

2.   What types of mathematical writing genres are students using in the tasks?

3.   What types of mathematical writing genres are missing?

4.   Create a task to provide students mathematical writing opportunities. Consider:

   a.   Why is this activity important for multilingual learners to experience?

   b.   What is the goal of the writing task?

   c.   What will students write about and to whom?

   d.   How will this specific writing task facilitate multilingual learners' development?

5.   Based on your analysis of the curriculum, what perceptions would students develop about writing in math?

## Reflect

- How would you explain to a colleague or parent why investing time to incorporate mathematical writing into your classroom benefits multilingual learners?

# CHAPTER 12
# ENHANCE CURRICULUM MATERIALS FOR MULTILINGUAL LEARNERS

## Key Concepts

In this chapter, you will

- ✓ analyze and enhance curriculum materials for multilingual learners.

- ✓ examine alternative mathematical conventions and representations.

- ✓ discuss the importance of analyzing, enhancing, and enacting curriculum materials to facilitate the mathematical learning of multilingual learners.

Ms. Bristow could not always anticipate when contexts in her mathematics curriculum materials would be problematic for her students. When she introduced a context involving postage stamps, she brought in letters and bills with postage stamps on them. She also brought sheets of stamps for her third graders to examine. She showed pictures of a post office, because some of the students in her class had never visited a post office or used postage stamps. Her students asked questions about how postage stamps work, such as "Are they like food stamps?" This incident gave her new insights into examining her curriculum materials (Chval & Chávez, 2012). Given students' varied cultural, educational, and life experiences, teachers must consider how their curriculum facilitates or restricts access to mathematics and language.

> *Given students' varied cultural, educational, and life experiences, teachers must consider how their curriculum facilitates or restricts access to mathematics and language.*

As stated in Chapter 5, this does not mean you should only use contexts your students already know since this will do little to prepare them for when they encounter unfamiliar contexts. Instead, you must be strategic in selecting, introducing, and enhancing contexts that can expand multilingual learners' understanding of the world.

As teachers of multilingual learners, we found curriculum materials often contained unfamiliar contexts and/or ambiguous or confusing language. Both characteristics created challenges for multilingual learners. In your examination of the use of visuals in print or online curriculum materials, you may have also noticed when visuals are absent, ambiguous, irrelevant, and/or unnecessary. Since there is no one curriculum that is relevant for all students (Gutstein, 2003), we regularly worked with teachers to develop new or enhance existing curriculum materials to challenge and engage multilingual learners. This led us to questions such as these: What should teachers attend to when analyzing and enhancing curriculum materials for multilingual learners? In what ways can curriculum materials be enhanced to increase access for multilingual learners?

## REFLECTING ON MATHEMATICS CURRICULUM

Read the two mathematics word problems in Try It! 12.1 and use questions a, b, and c to guide your analysis and revision of the problems. Later in the chapter, we provide ideas for how to conscientiously craft word problems with multilingual learners in mind.

---

### ⭐ Try It! 12.1

1. Madeline was changing subway trains and dropped some coins. She now had ninety-five cents. If she had $2.56 before losing the money, how much change did she lose?

2. Madeline has $2.56. She buys some apples. She has $0.95 left. How much did the apples cost?

(continued)

(continued)

> For each problem:
>
> a.    Identify features that may facilitate access for multilingual learners.
>
> b.    Identify features that may be problematic for multilingual learners.
>
> c.    Revise the problems to address the problematic aspects you identified.

We have often heard statements such as "Word problems should be 'short and easy' for multilingual learners" or "Word problems should not be given to multilingual learners." Eliminating access to word problems is not the answer. Lowering mathematical expectations for multilingual learners is not the answer. In Try It 12.1, in the second problem you may have noticed the use of numerals, present tense, shorter sentences, and the same mathematical units ($ rather than $ and *cents*). You may have also noticed some problematic features in the first problem, such as the use of

- different tenses;
- a conditional clause;
- *changing* and *change* in the same problem;
- a context that may not be familiar (i.e., changing subway trains);
- longer sentences;
- *ninety-five* rather than 95;
- *dropped* and *lose*;
- "some" coins; and
- words that have multiple meanings (i.e., *change*).

It is important to recognize these subtleties that may influence multilingual learners' success.

## WHAT THE RESEARCH SAYS ABOUT CURRICULUM ENHANCEMENT FOR MULTILINGUAL LEARNERS

We define curriculum as the "instructional materials, activities, tasks, units, lessons, and assessment" (National Council of Teachers of Mathematics, 2014, p. 70) that support students to meet mathematical content standards. Thus, mathematics curriculum provides the foundation for classroom interactions and is inextricably intertwined with learning opportunities provided to students. However, "the design process for mathematics curriculum materials has not involved sufficient attention to language diversity and creating tasks and contexts that facilitate the participation of multilingual learners" (Chval, 2010,

p. 15). As a result, multilingual learners could encounter barriers as they work with curriculum materials (Doerr & Chandler-Olcott, 2009; Gottlieb & Ernst-Slavit, 2019). Therefore, teachers are in a position that requires them to analyze, enhance, and enact curriculum materials to facilitate the mathematical learning of multilingual learners.

*Teachers are in a position that requires them to analyze, enhance, and enact curriculum materials to facilitate the mathematical learning of multilingual learners.*

Some specific challenges that multilingual learners face when learning to read material in English is syntax, or the word order in a sentence (Haynes, 2005), particularly in word problems (Gottlieb & Ernst-Slavit, 2019). This order, or structure, determines certain meanings. "Sentences in mathematics convey complex relationships and abstract ideas. To achieve this, they often have linguistically complex sentence structures with dependent clauses" (Adoniou & Qing, 2014, p. 7).

## Strategies for Crafting Language to Better Support Multilingual Learners

Linguistic enhancements (adapted from Abedi & Lord, 2001, p. 221) to support multilingual learners in the mathematics classroom include the following:

▶ Change from passive to active voice. For example, change "5 cookies are shared by Juan with his friends" to "Juan shares 5 cookies with his friends."

▶ Move from long to short nominal phrases (i.e., phrases that act as nouns). For example, change "The top speeds for different cats are shown in the graph. Describe the shape of the top speeds graph" to "The graph shows top speeds. Describe the shape of the graph."

▶ Separate sentences when possible. Replace a conditional clause with two separate sentences. For example, change "A platter holds 10 cups of strawberries, with each cup having 65 calories in it" to "A platter holds 10 cups of strawberries. Each cup has 65 calories."

▶ Change obscure question phrases to straightforward question phrases. For example, change "What amount of the cooked pasta is left after Alex eats?" to "What fraction of the pasta is left?"

▶ Replace complex verb forms with present tense verbs. For example, change "A student *has been walking* 1 mile home from school" to "A student *walks* 1 mile each day."

▶ The absence of personal pronouns (*I*, *you*, *she*, *he*, etc.) may obscure the presence of human beings in the text and create a distance with the student (Herbel-Eisenmann, 2007). Change word problems with inanimate or abstract objects to pronouns or animate nouns. For example, change "90% of test scores were passing" to "90% of students passed the math test."

## ANALYZING CURRICULUM MATERIALS FOR MULTILINGUAL LEARNERS

Ideas that we have discussed in earlier chapters are important considerations as you examine your curriculum materials. For example, the following questions can guide your efforts:

▶ Are multilingual learners familiar with the context of the curriculum materials? (Refer back to Chapter 5.) If not, how can I situate the mathematical concept in a context with which multilingual learners are familiar? Alternatively, how can I introduce this context to the students?

▶ Do visuals in the materials contribute to multilingual learners' mathematical learning? (Refer back to Chapter 6.) If not, what visuals can I use to contribute to multilingual learners' mathematical learning?

▶ Are multilingual learners familiar with the language? (Refer back to Chapter 8.) If not, how can I build a shared and common meaning for unknown terms?

▶ Are instructions clearly and simply stated? If not, in what ways can I adapt the instructions to make them more accessible for multilingual learners?

Use these questions to analyze the mathematics handout in Try It! 12.2.

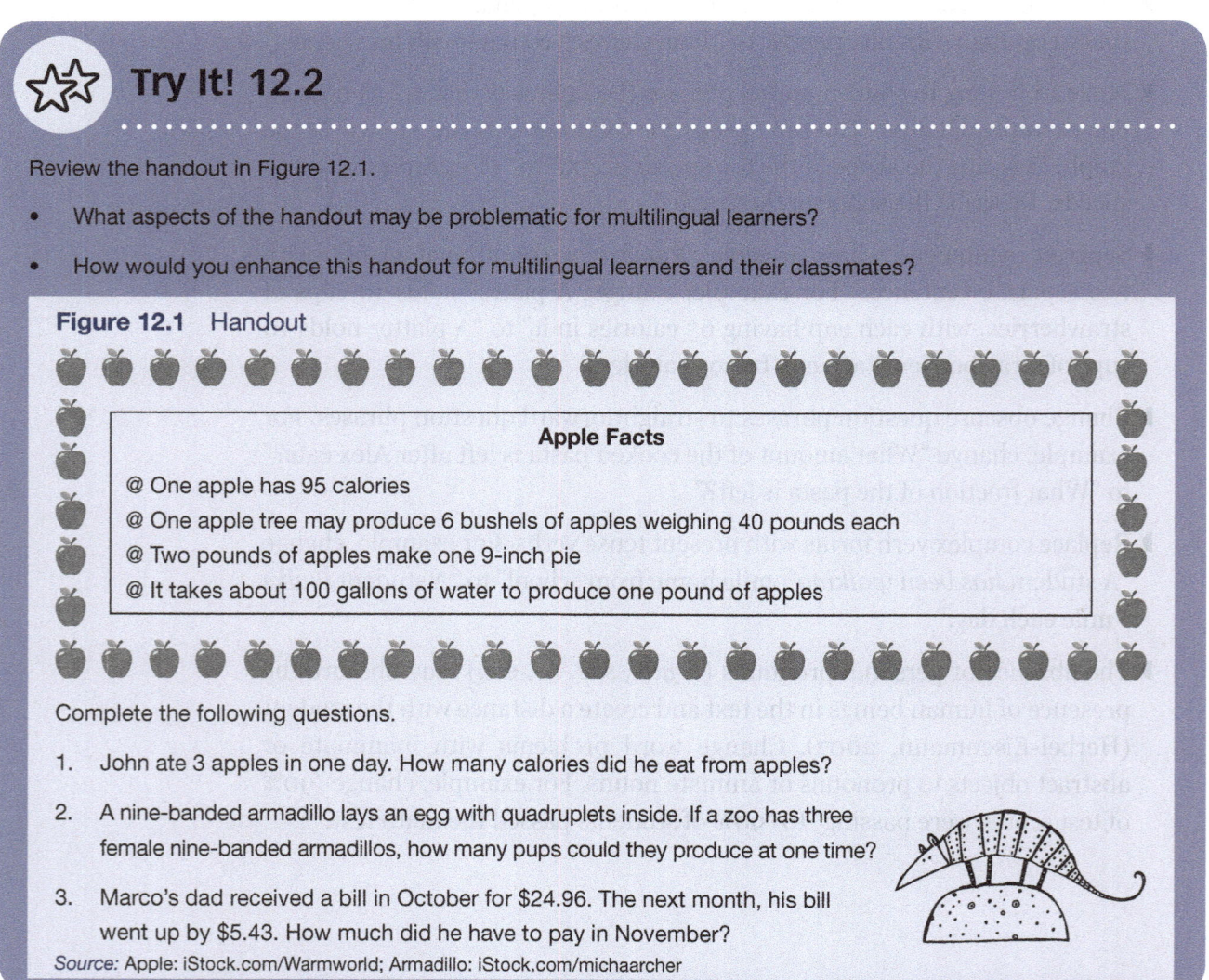

### ⭐ Try It! 12.2

Review the handout in Figure 12.1.

- What aspects of the handout may be problematic for multilingual learners?

- How would you enhance this handout for multilingual learners and their classmates?

**Figure 12.1**    Handout

**Apple Facts**

@ One apple has 95 calories
@ One apple tree may produce 6 bushels of apples weighing 40 pounds each
@ Two pounds of apples make one 9-inch pie
@ It takes about 100 gallons of water to produce one pound of apples

Complete the following questions.

1. John ate 3 apples in one day. How many calories did he eat from apples?

2. A nine-banded armadillo lays an egg with quadruplets inside. If a zoo has three female nine-banded armadillos, how many pups could they produce at one time?

3. Marco's dad received a bill in October for $24.96. The next month, his bill went up by $5.43. How much did he have to pay in November?

*Source:* Apple: iStock.com/Warmworld; Armadillo: iStock.com/michaarcher

4.  Martha's brother has given her five CDs. Her friend Annie has given her three CDs. If she already had ten CDs, how many CDs does Martha now have?

5.  The Hollywood Walk of Fame has a total length of 15,108 feet. Josephina measured that each star had an interval of six feet. How many stars are there on the walk?

6.  Do you remember? Solve.

1.  $3\frac{3}{5} + 1\frac{1}{5} =$          3.  $\frac{8}{9} \times 4 =$          5.  $9 \times \frac{4}{4} =$

2.  $7\frac{8}{9} - 4\frac{2}{9} =$          4.  $\frac{3}{7} \times 0 =$          6.  $568 \times 12 =$

As you analyzed the handout in Figure 12.1 using the guiding questions, you may have identified the contexts, visuals, language, and instructions listed in Figure 12.2.

**Figure 12.2**   Contexts, Visuals, Language, and Instructions in Handout (Figure 12.1)

| Contexts | • Number of contexts |
| | • Outdated technology—out-of-date references (e.g., CDs) |
| | • Unfamiliar contexts |
| Visuals | • Use of images and their location on the handout |
| | • Use of phrasal verbs (i.e., *went up*) |
| Language | • Words may be unfamiliar |
| | • Grammatical structures (e.g., use of verb tenses) |
| | • Numbers vs. numerals |
| | • Unnecessary information |
| | • Use of conditional clause (i.e., *if*) |
| | • Length of some sentences |
| Instructions | • "Complete" and "Do you remember?" may confuse learners |

## RESEARCH RECOMMENDATIONS FOR ENACTING MATHEMATICS CURRICULUM

Teachers need the autonomy to analyze when curriculum and instruction require adaptations and enhancements to provide access and opportunity to multilingual learners (Garcia & Gonzalez, 1995). According to Chval (2010), "teachers must consider how to enhance and enact curriculum materials for multilingual learners as these students spend most of their time in mainstream classrooms" (p. 107).

## Strategies for Scaffolding Curriculum Materials to Support Multilingual Learners

Some additional recommendations related to analyzing, enhancing, and enacting curriculum materials for multilingual learners include the following:

▶ *Build meaning, adapt materials, and include significant contexts.* Providing a vocabulary list or definitions of mathematical terms is not enough to build the meaning necessary to solve a mathematical problem (Vomvoridi-Ivanović & Chval, 2014). When teachers find challenging language or contexts for multilingual learners in their curriculum, they can either build meaning through sustained activities (see Chapter 8; for additional reference see Chval, Pinnow, & Thomas, 2015; Vomvoridi-Ivanović & Chval, 2014) or adapt the curriculum materials to include significant contexts for all the students in class (see Chapter 5).

▶ *Use a thematic approach* with the same context to provide a series of related tasks over multiple lessons at the beginning of the year. Then, *add more contexts over time* (Pitvorec, Willey, & Khisty, 2011).

▶ *Use multiple representations* in the mathematics classroom to positively impact multilingual learners' mathematical learning and language development (Moschkovich, 2013, 2015). Multiple representations, such as student drawings and graphs, physical objects, contexts, and symbols, can be used by multilingual learners to expand mathematical thinking and provide alternative ways of communicating ideas. By introducing these tools to students as they work to understand mathematical concepts, ideas, and content, teachers provide "linguistic and intellectual support" (Anhalt & Ondrus, 2011, p. 202).

▶ *Prepare students to succeed with typical texts in mathematics.* According to Moschkovich (2013),

> Typical written texts in mathematics include not only word problems and mathematics textbooks, but also other students' written explanations that are shared in small groups and a teacher's or a student's solution written on the board. Typical written texts also include assessment problems and scenarios for modeling. Oral texts include explanations, descriptions of solutions, conjectures, and justifications. (p. 54)

## RECOGNIZING DIFFERENT MATHEMATICAL CONVENTIONS AND REPRESENTATIONS

Another important aspect of mathematics curriculum includes mathematical conventions and representations. Multilingual learners bring different background knowledge and resources to the classroom (Moschkovich, 2002). For example, a multilingual learner may use conventions learned in prior schooling experiences or from a parent that are unfamiliar to the teacher. As a result, multilingual learners may not recognize the use of specific symbols in curriculum developed in the United States. For example, consider Figure 12.3.

*A multilingual learner may use conventions learned in prior schooling experiences or from a parent that are unfamiliar to the teacher.*

**Figure 12.3**  Mathematical Conventions

|  4°  |  3°  |
|------|------|
| 0,20 | 1,60 |

- What is the meaning behind these representations?

- How does the meaning of each representation differ across cultures?

In Latin America, the numbers 4° and 3° are used to express Celsius *and* ordinal numbers. Students who studied in Latin America may be familiar with "3° grade" to mean "3rd grade." Furthermore, in some Spanish-speaking countries and some Middle Eastern countries, comma notation is used for time, money, and measurement and can act like a decimal. One would read 0,20 as "zero comma twenty" or "comma-twenty." As a result, teachers should not assume that all cultures share the same conventions, and provide support to their students accordingly (Solano-Flores, 2011).

Moreover, multilingual learners who learned mathematical conventions or algorithms in other countries may use different approaches to calculations (Kersaint, Thompson, & Petkova, 2014). They may write mathematical representations and numerals in different ways from students who have been introduced to them in the United States. For example, you could encounter a student who wrote a division problem, 1,509 ÷ 26, as shown in Figure 12.4.

**Figure 12.4**  Example of Long Division Algorithm

```
  1509 | 26
– 130   58
 ─────
   209
 – 208
 ─────
     1
```

Although the placement of the divisor and dividend could be considered the opposite of the long division algorithm taught in the United States, the computation steps are the same. In contrast to the method taught in the United States, this method keeps track of the steps on the left-hand side (as opposed to the right-hand side). Then, the result from each step of the algorithm is written underneath the divisor (like the U.S. version). The remainder of 1 is shown in blue at the bottom.

**VIDEO 12.1:**

Lusto, C. (2012, January).
*Japanese multiplication* [Video].
YouTube. https://www.youtube.
com/watch?v=85Vd0NpL32k

Watch Video 12.1, *Japanese Multiplication* (Lusto, 2012), stopping at 5:25.

- Were you familiar with the multiplication method explained in the video? What other international methods of calculation do you know?

- What strategies could you use when you encounter other international methods of calculation in your classroom?

Your students may be interested in exploring different algorithms across cultures. Check out the Algorithm Collection Project online. When students understand why algorithms work and the meaning behind the symbols, they are more successful with them. In addition, this resource can support your own practice. We have received inquiries from teachers about students who move to the United States and use different algorithms and symbol systems. Consider the exchange with one teacher in Transcript 12.1.

**Transcript 12.1**

| | |
|---|---|
| Teacher: | I have been teaching the U.S. division algorithm in my class. I have some multilingual students in my classroom from Europe and Asia who use different algorithms. I don't understand how they work. Should I make them do it the U.S. way? |
| Our response: | Do they get the correct answer? |
| Teacher: | Yes, every time. |
| Our response: | Do they do the problems efficiently, or does it take them a long time? |
| Teacher: | They are always the first students done with the problems. |
| Our response: | Ask them to teach you their method. This will validate their approach. As you teach them the U.S. algorithm, ask them to compare the methods. What is similar? Different? Why do both work? |

The resources on the Algorithm Collection Project site, such as the TODOS *Mathematical Notation Comparisons Between U.S. and Latin American Countries* (Lopez, n.d.) and *Long Division Algorithms Collected in the European Union* (2003), will provide you with necessary background on how algorithms work and what to look for when analyzing your multilingual learners' work (Orey, 1999).

## THINKING ABOUT MATHEMATICAL CONVENTIONS AND REPRESENTATIONS IN YOUR PRACTICE

Curriculum materials can foster standardized procedures and habits of mind (Hirsch, Lappan, & Reys, 2012). Students who come from diverse backgrounds may be familiar with different procedures, forms of calculation, or habits of mind. In these cases, focus on the mathematical thinking performed by the student, not on the correct use of a determined procedure. Multilingual learners who know other procedural approaches should have opportunities to use their prior knowledge. Students who are not allowed this opportunity may stop participating in group or whole-class discussions as their prior knowledge is devalued (Liu, 2015). You can prompt students to explain

▶ how and why an algorithm works;

▶ how different algorithms are connected; and

▶ which strategy/algorithm is more efficient, and why.

You can support students whose families know different methods by creating a chart comparing different conventions. If, for example, the student's calculation method is efficient and demonstrates understanding, acknowledge this in front of the student. If the student still can learn more efficient habits of mind or methods of calculation, the student's background knowledge should be used as a positive base to build new understandings. You also have to consider the curriculum tasks that you send home. Parents may report feelings of frustration with algorithms students are using when they differ from those with which they are familiar. Parents may also experience frustration if extended family in their country of origin have children of the same age, but are learning mathematics curriculum that is more advanced. We will discuss parent engagement in Chapter 13.

## THINKING ABOUT CURRICULUM ENHANCEMENT IN YOUR PRACTICE

You have likely noticed situations where all of your students, including multilingual learners, have been confused by curriculum materials. For example, multilingual learners may not be familiar with specific set phrases (e.g., *once in a while*); phrasal verbs (e.g., *ask for*, *come up*, *look after*); and idioms (e.g., *a piece of cake*, *a penny for your thoughts*, *break a leg*; Folse, 2004), as discussed in Chapter 9. Alternatively, multilingual learners may be unfamiliar with particular contexts as discussed in Chapter 5, visuals as discussed in Chapter 6, or conventions as discussed in this chapter.

For Try It! 12.3, pick one lesson, task, or activity with multiple contexts from your mathematics curriculum materials.

 **Try It! 12.3**

Analyze the contexts, language, visuals, and instructions in the lesson, task, or activity you chose. Determine which aspects should be enhanced for multilingual learners. Discuss your findings with a colleague.

## Reflect

- What will you consider when enhancing curriculum for multilingual learners?

- How will you approach different mathematical algorithms and conventions from different countries?

# CHAPTER 13
# ENGAGE WITH PARENTS AND FAMILIES OF MULTILINGUAL LEARNERS

## Key Concepts

In this chapter, you will

- ✓ understand different multicultural parenting styles and their influence on parental engagement at school.

- ✓ identify ways to foster multilingual parent and family engagement in the mathematics classroom.

Moving to another country is not easy for children or their parents. Beyond language, families face multiple cultural challenges in these transitions, especially adapting to new school systems and cultures. You may have encountered this type of experience or interacted with families who transitioned to homes in different countries or even within the United States. In this chapter, you will explore how different parenting styles and cultural values may influence the ways parents engage in their children's education. Specifically, we address some conceptions that could affect the relationship between multilingual parents and personnel in their children's school. Then, we focus on alternatives to foster multilingual parent engagement in the classroom.

## REFLECTING ON YOUR EXPERIENCES

Imagine that you are a parent who had to move to another country alone with your eight-year-old daughter. After three years in the new country, you can speak the new language for daily life and basic tasks at work, but you prefer to speak English at home and in other contexts. In contrast, your daughter has mastered the second language and prefers to speak it instead of English. She understands when you talk to her, but refuses to speak in English. She is now in third grade, and you are finding it difficult to support her with her homework.

> ▶ What challenges might you encounter when trying to support your daughter at home?

> ▶ How can the teacher or other school personnel support you as you face these challenges?

Some of the challenges you may have considered include communication barriers; emotional responses, such as frustration; and cultural identity conflicts. School personnel and community members can support multilingual and immigrant families during their transitions to a new culture.

## WHAT THE RESEARCH SAYS ABOUT MULTILINGUAL PARENT AND FAMILY ENGAGEMENT

Many families decide to migrate because of financial necessity or political climate in their countries of origin (Crawford & Dorner, 2019; Dettlaff, 2008), while other families travel to the United States to study or work. In any case, these families could experience fear, stress, isolation, loss, and uncertainty in the United States. They have to adapt to a different culture where they may not have the established support systems they had at home, such as friends and family. Families may be unfamiliar with the U.S. school system and may not know how to negotiate it (Arias & Morillo-Campbell, 2008; Castro et al., 2015; WIDA Consortium, 2015; Wilder, 2014). Small things, such as who is responsible for what, whom to ask when they have a question, or even where to park when visiting the school, may be stressors for immigrant families who are adapting to a new life in a different country or state.

Culturally sustaining practices that enhance their participation and engagement at schools—particularly in mathematics classes—are required, since parental engagement is positively related to children's achievement in school (Civil & Menendez, 2010). According to the WIDA Consortium (2015), "*traditional* forms of family engagement often focus on parent attendance at important school events (e.g., parent-teacher conferences), parents volunteering in the school, or parents helping children with homework" (p. 2). They also include resources for parents such as translation services, workshops, and cultural activities. *Nontraditional* forms of family engagement, in contrast, include "supporting parents from diverse backgrounds as they build leadership and advocacy skills, offering opportunities for parents to learn about their rights within the U.S. school system, or offering opportunities for staff to build the skills and knowledge needed to successfully engage with culturally and linguistically diverse families" (WIDA Consortium, 2015, p. 2; see also Bonney, Dorner, Trigos-Carrillo, Song, & Kim, 2019).

As you reflect on the ways you work with multilingual parents and families, think broadly about other people in your school or school district who can be part of this conversation. For example, take into account what administrators and other stakeholders can do to support multilingual parent engagement. Finally, consider specific actions you can carry out in your classroom to support multilingual parents and families.

## MULTICULTURAL PARENTING STYLES

Too often there is a mismatch between U.S. teachers' and multilingual parents' expectations of one another's roles in children's education (Rodríguez-Brown, 2010). For example, for many Latinx families, there is a difference between the expressions "to educate" and "to teach." *Educating* in Spanish "encompasses moral and ethical values and social behaviors. Latino parents see their role as contributing actively to the *educación* of their children, whereas they may see teaching (*enseñar*) as pertaining to teachers and schools (Valdés 1996; Zarate 2007)" (Civil & Menendez, 2010, p. 2). In one study, Valdés (1996) found that for some multilingual parents the moral education of their children is their primary responsibility. To *educate* their children's attitudes and behaviors, parents frequently use *consejos*, spontaneous advice homilies, to influence children's behavior. Thus, teachers must understand and be attentive to the diverse cultural expectations that multilingual parents bring to school, highlight the community cultural wealth of multilingual families (Yosso, 2005), and adopt nontraditional forms of parent engagement (Dorner, Song, Kim, & Trigos-Carrillo, 2019).

### Cultural Values

For Try It! 13.1, consider your family values (i.e., what is important in life) and how they may or may not differ from those of multilingual families.

## ⭐ Try It! 13.1

Create a list of five core values that are important in your family.

▶

▶

▶

▶

▶

Now, rank the three most important values for you. Write your three core values in the boxes in Figure 13.1.

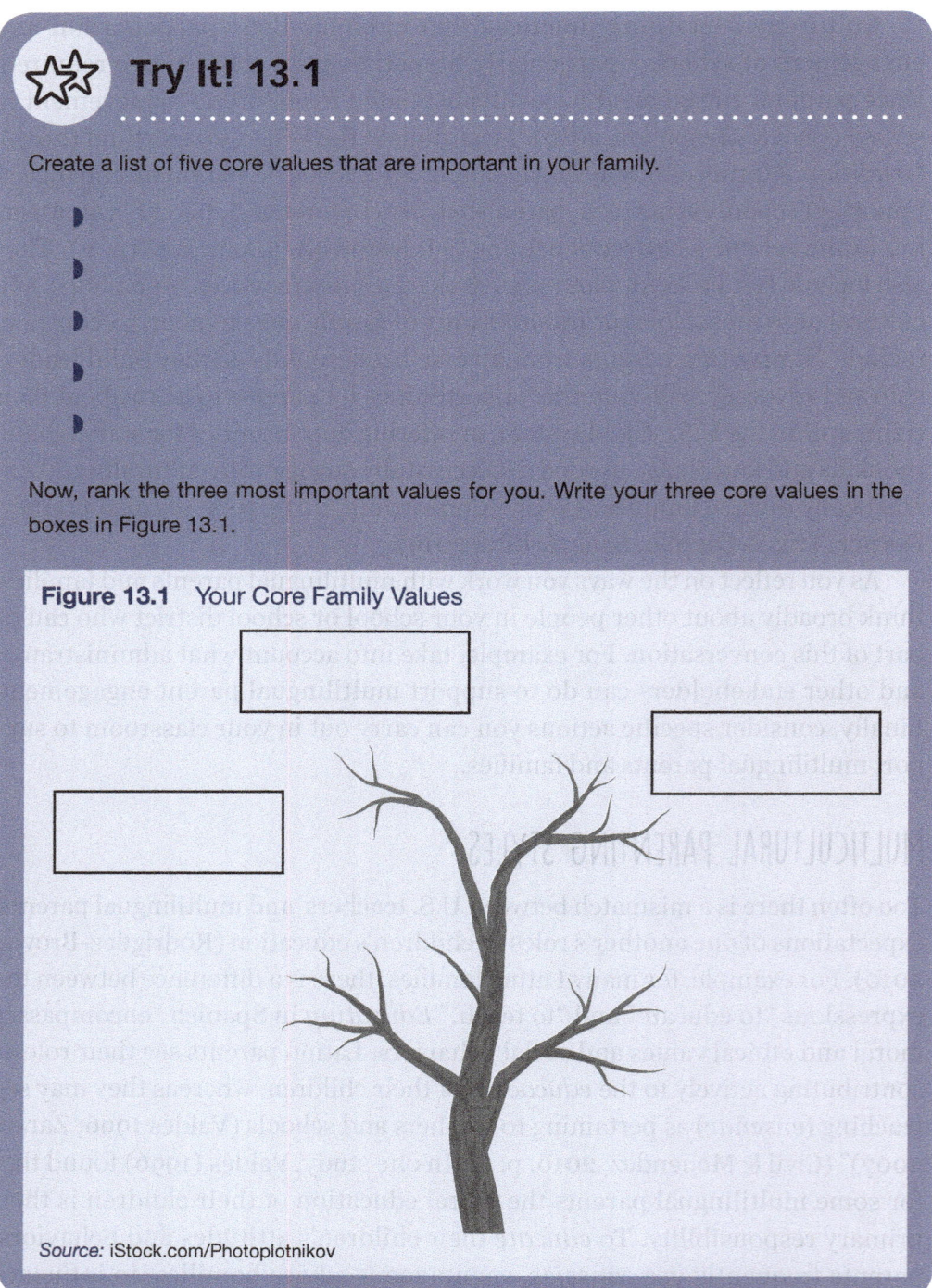

**Figure 13.1**    Your Core Family Values

*Source:* iStock.com/Photoplotnikov

We surveyed teachers, and many of them identified the following values:

▶ Work ethic

▶ Individualism

▶ Family

▶ Education

▶ Success (many times measured by wealth)

▶ Performance

‣ Independence

‣ Perseverance

We are aware that these values do not represent the aspirations of every U.S. family because family values vary based on individuals' sociocultural background and socioeconomic status. However, this exercise helps us understand multicultural differences and become aware of different approaches to education. Now consider how your family values compare to the values of one multilingual mother.

Lorena is a woman from Colombia, who has lived in the United States for five years. When she arrived in the United States with her family, her son was three years old; now he is eight. Read Lorena's statement about the way she educates her son:

> **66** *Before my eight-year-old son goes to school, I make sure he has a good breakfast, is clean, and is ready to attend school. I don't like when he forgets something he would need during the day or is not well dressed. Beyond that, my husband and I consider it very important that he is a well-educated child. That means that he is collaborative, he is kind to others, he treats people with respect, and he is a good person. Even though we attended graduate school, we do not care as much about the content he learns at school as we care about habits and attitudes that will stay with him during his lifetime. Positive daily attitudes include greeting people respectfully, following instructions, not using "curse" words, being careful with others, and practicing good manners. My daily advice or* consejos *to him are: "Always be a good boy," "Don't be rude or unfair to others," "Treat others with respect, particularly the elderly or* los abuelos," "Don't lie," *and "Be* cortés" *(being courteous or polite means holding the door for others, offering help if needed, greeting, and waiting for his turn).* **99**

## STOP AND THINK

Stop and think about Lorena's statement.

• What are the most important values for Lorena's family?

• How does her view of her son's education differ (or not) from your own views of education?

You may have noticed that Lorena's family focuses on attitudes and character development. Respect is very important for her, and she focuses on instilling positive values in her son. Now let's learn from another mother, Rosa, who moved with her family from Los Angeles, California, to a small town in the Midwest in 2005. Although Rosa and her family were from Mexico and spoke Spanish at home, Leo, her third-grade son, had never been to Mexico. The interview in Transcript 13.1 was conducted in Spanish and then translated to English.

**Transcript 13.1**

| | |
|---|---|
| Interviewer: | When you think of your child, what is it that makes you feel most proud of him? |
| Rosa: | Oh! Everything he does. I feel so proud of my son, more so when he brings me good marks or that at school they tell me that he is progressing more—it makes me proud of him. |
| Interviewer: | Yes, how good. And what is it that makes you smile about your son—what makes you happy about him? |
| Rosa: | His heart. |
| Interviewer: | Really? How good! |
| Rosa: | The heart that my son has—he has a very good heart . . . . |
| Interviewer: | And you have lived in different parts of the States or only in California? |
| Rosa: | No! I have lived in Los Angeles. I have lived in San Francisco, in Arkansas, and here. |
| Interviewer: | How nice, so you have been able to know different parts, and 13 years is a good amount. |
| Rosa: | 13 years, and ever since I haven't been back [to Mexico], not once. |
| Interviewer: | You didn't go back, and do you miss it? |
| Rosa: | Yes, of course! |
| Interviewer: | Family? |
| Rosa: | My family, the food, the traditions . . . . He [Leo] tells me, "Mommy," he comes and tells me, "Today a new girl has arrived," or "Mommy, there is a girl that I like." Or whatever little thing or "Look, Mom, I made a drawing for a girl. What do you think? Is it beautiful or not?" He is like that. He comes home each day, and later now when he arrives and he doesn't tell me anything, I will ask him, "Leo, what did you do in school?" "Nothing, Mommy. Today, nothing happened. Everything normal." . . . He expresses himself without fear that he is going to be reprimanded. When he does something that he is not supposed to do, then, of course, I let him know that he shouldn't be doing that. Or also how to express himself with other people. I teach him to be respectful to the elderly, to his teachers, to people older than him . . . . He makes his bed. Since he was very little, he makes his bed. He looks for his clothes. I only wash it and I put it in his room and I tell him, "Leo, here are your clothes. Can you put them away?" "Okay, Mommy." He knows where everything goes—the sports shoes, the T-shirts, the trousers. |
| Interviewer: | He is very organized. |
| Rosa: | Yes, and when the twins [referencing Leo's younger siblings] do something—they go to his room to find something—he says, "No. Don't come to my room. If you need something, come to me and ask me." Because I have taught them that way. They are accustomed to respect each other's space. Because the twins have their own room, the girl has hers, and they respect my room very much. He knows that he can't come to my room if he doesn't have permission. And if he wants to take something from my room that I have that will be useful to him, he comes and asks me, "Mommy, can I use this from your room?" I say, "Okay, but come and show me what you took." |
| Interviewer: | Respect. |

| Rosa: | And I also teach the twins that he is the older brother and they have to respect him. |
| Interviewer: | How lovely. |
| Rosa: | And he tells them, "I am the older brother, and you have to respect me and you have to be respectful to me." And I tell him, "It's good for them to respect you, but you also have to respect them. If you want respect, you have to earn it." |
| Interviewer: | Exactly. |
| Rosa: | Respect won't be given to you because you say that you are the man of the house. And you also have to respect. You earn respect. Respect has to always be earned. If you want respect from your brothers, you have to respect their things. Because he happens to forget, [and he thinks] "just because I am the older son I am going to do what I want." And the twins confront him. "You have to respect us!" the twins tell him. "You also have to respect," I tell him. "You have to have respect. If you want respect, you have to have it." |

## STOP AND THINK

Stop and think about Rosa's interview in Transcript 13.1.

- What values are important for Rosa?

- What does she want to know about her son when he comes home from school?

- What does she want to teach to her son?

In a study about core values among Latinx families, Civil and Menendez (2010) found that the most important values for them are family (*familia*), respect (*respeto*), and trust (*confianza*), as shown in Figure 13.2. The value of family is central to Latinx families, and it includes extended family. Respect is sometimes related to maintaining one's roles, and trust requires building a relationship with families. These ideas extend to other multilingual families as they may have some core values and beliefs about education that you may not have considered or that may be different from yours. Learning from multilingual families' culture is important to understand their approach to education, and how you can create productive partnerships with them.

The value of family is central to Latinx families—like other multilingual families—because in Latin America, people traditionally live near their extended family, so they often become a close community. For that reason, Latinx families collaborate with and receive support from close relatives. In terms of respect, this value is strongly associated with courtesy rules. From the time children are little, they are expected to follow good manners and courtesy norms that are particular to their culture, especially with elders, who tend to live with their families. Finally, because of their strong sense of community, trust is highly valued among Latinx families. Of course, these are representative family values among Latinx, but in Latin America, as in other places, there

**Figure 13.2**    Core Values of Latinx Families

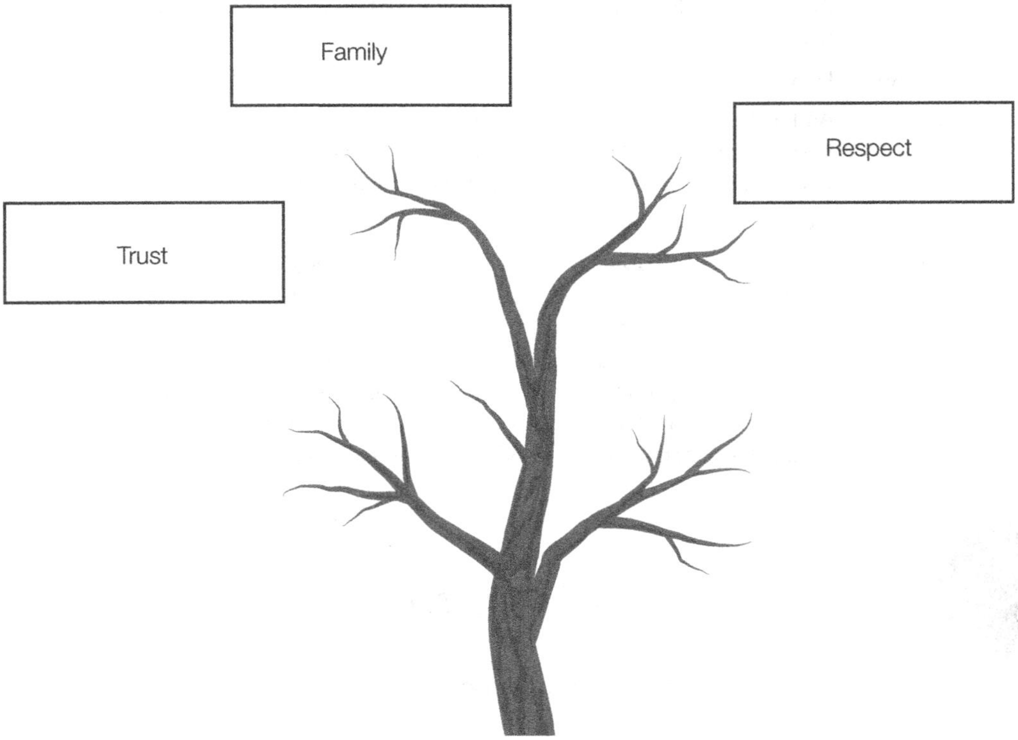

Family

Respect

Trust

*Source:* iStock.com/Photoplotnikov

are many cultural differences among countries, and it is difficult to generalize. Even within countries there are cultural differences; people who live on the coasts are different from people who live in the mountains. People from different socioeconomic backgrounds may also have different values. Every multilingual family has a set of core values and appreciates education in different ways.

It is important to consider how cultural differences may affect the ways parents and families engage with their children's school and education, and how you can develop a strong relationship with multilingual families and communities. For example, some multilingual families may view the teachers and school staff as academic experts and not want to encroach on their expertise out of respect (Arias & Morillo-Campbell, 2008). In some countries, teachers are highly respected and valued. For this reason, some parents may not be accustomed to asking questions about their children's academic progress. Being aware of these differences and learning directly from multilingual families can help you build stronger relationships. Let's look at how Ms. Martínez fosters the value of family in her classroom.

*It is important to consider how cultural differences may affect the ways parents and families engage with their children's school and education, and how you can develop a strong relationship with multilingual families and communities.*

## Emphasizing the Value of Family

Ms. Martínez recognized the importance of family and respect, and emphasized it in her classroom. As you read Transcript 13.2 (first published in Chval, 2012, pp. 80–81), consider how she incorporates the values of family and respect into her mathematics classroom.

**Transcript 13.2**

| Ms. Martínez: | You care about your family. Is there anyone here who does not care about their family? [*No response*] Ah, everybody cares about their family. Do you care about school? |
|---|---|
| Students: | Yeah. |
| Ms. Martínez: | Do you? How do you show that you care about school? Violetta is annoyed. I can see it in her face. Hold on a second. Her face is all crunched up, and I know she wants to listen, and she's reacting to something that Matthew is doing. Is Matthew showing us that he cares? |
| Students: | No. |
| Ms. Martínez: | Is Violetta hearing what all is going on? |
| Students: | No. |
| Ms. Martínez: | No, she's busy taking care of a problem here. And Matthew is not listening very well. Matthew, we care about you, so what can we do to help you out? Because this is a family right here. We spend lots and lots of hours together, don't we? |
| Matthew: | Yes. |
| Ms. Martínez: | So, we are a family. When you do something, it affects all of us, doesn't it? All of us, right now, are paying attention to you, because you cannot pay attention. We have to pay attention because we care about you. Do you care about us? |
| Matthew: | Yeah. |
| Ms. Martínez: | How are you going to show us that you care about us? |

Ms. Martínez went on to say:

> " *If I show you respect, I expect respect right in return. If I work very, very hard, I expect the same from you—that you work very, very hard. If I listen to you, then I want you to listen to me. I show you that I care by the work that I do. You show me that you care by the work that you do. That's how I want you to show me respect. How do you want me to show you that I care? Alejandro? Is there anything special that I should do, or am I already doing what you think I should? (Chval, 2012, p. 80)* "

*Source:* Reprinted with permission from *Beyond good teaching: Advancing mathematics education for ELLs*, copyright 2012, by the National Council of Teachers of Mathematics. All rights reserved.

# STOP AND THINK

Stop and think about Transcript 13.2 and Mrs. Martínez's statement.

- What did you notice about Ms. Martínez's classroom approach?

Ms. Martínez established certain expectations and characteristics related to what it means to live in a family, such as caring about each other, respecting others, and hard work. This is very important because Ms. Martínez is building a community with an established set of norms and expectations that aligns to her students' cultural norms and expectations. For multilingual or immigrant students, it is important that they feel cared for and belong to the classroom community (Orellana, 2016).

## Examining Teachers' and Multilingual Parents' Expectations

When we participate in school, either as teachers or as parents, we have expectations about our roles and about other people's roles in education. We act based on those expectations and sometimes judge others based on what we think they should be doing. As noted in Chapter 2, our expectations about others' roles can influence how we position them and thus how we interact with them.

For Try It! 13.2, examine your own expectations about parents' roles.

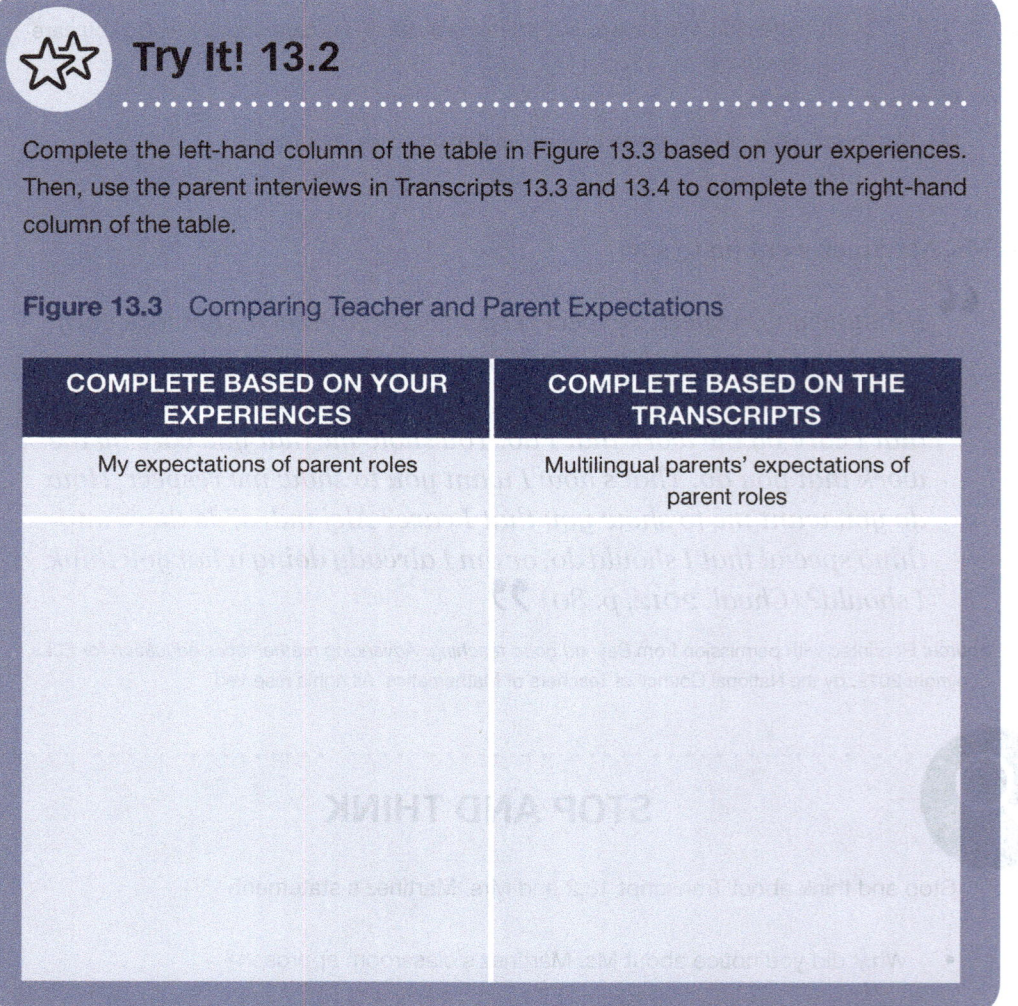

### Try It! 13.2

Complete the left-hand column of the table in Figure 13.3 based on your experiences. Then, use the parent interviews in Transcripts 13.3 and 13.4 to complete the right-hand column of the table.

**Figure 13.3** Comparing Teacher and Parent Expectations

| COMPLETE BASED ON YOUR EXPERIENCES | COMPLETE BASED ON THE TRANSCRIPTS |
|---|---|
| My expectations of parent roles | Multilingual parents' expectations of parent roles |
| | |

## Transcript 13.3

Adriana was born in Minnesota and moved to a small town in the Midwest in 2006. Although Adriana's family is from Mexico, Adriana does not speak Spanish. The following interview with Margarita, Adriana's mother, was conducted in Spanish and then translated to English.

| | |
|---|---|
| Margarita: | The truth is that my daughter is having a hard time with math. For example, yesterday we were looking at grades, and because I am not good with English and she doesn't speak good Spanish, we don't understand each other very well and I don't know how to explain things to her very well. I can't do it. I can't explain things to her, and she comes up with numbers where I don't know where she gets them from. |
| Interviewer: | So, you're saying that she makes up numbers? |
| Margarita: | Yes. So, for example, one time there was 1,005 times 8, but what she does with that, she multiplies 8 four times, which is 32. |
| Interviewer: | So, she multiplies 8 by 4, but she doesn't need to do that. |
| Margarita: | Yes, 8 multiplied by 0 is 0 plus 4 is 4. She doesn't want to listen to me, she doesn't want to pay attention to me, and that's how I struggle with the children. When they are smaller, it's better, but when they grow a little older, they don't pay any attention to me and they tell me, "You don't know. My teacher knows." . . . |
| Interviewer: | Did any of her friends leave? |
| Margarita: | No, the one that left was her. She went to X school and I didn't like it there—I didn't like it. I felt that they treated us very badly . . . . It's apparently the best school academically here, but I didn't like the treatment. When we were at the open house, we went there for the teacher to take care of us, but we waited a long time and nothing. We were the only Hispanic people there that day . . . . The other school [referring to the same school] was where she didn't want to go, and she always likes to go to school, and she would say, "My teacher shouts at me very much." |
| Interviewer: | In the other school? |
| Margarita: | Yes. The teacher said to her, "If you don't like it here, stay at home." And we wanted to change teachers, and they told me, "No, we don't do that." And there was a problem, and the principal asked me and said, "She's the best teacher we have," and I thought to myself, "If she's the best, imagine what the worst would be like," and I told her [Adriana], "We can change schools if you want to." |

## Transcript 13.4

| | |
|---|---|
| Rosa, Leo's Mother: | If I am not calm about something in relation to my children, I will communicate it to them [school]. I even have very beautiful communication with the teachers. They always tell me if I have any questions or some worries about your son, I will tell them. When I see my children improving, it is okay. And if I see them declining this year, I ask "Why, why?" to find out if the child is paying attention or if he is having problems. So I go to speak to them . . . . I have seen, in the conferences, that there are parents that go and the teachers ask them, "Do you have any questions?" They reply, "No, everything is okay." And I am not like that; I am always concerned about all my children's matters. To me, I care so much about my children's things. When it |

(continued)

(continued)

|  | comes to my children, it concerns me a lot, and I want to know why they go down and when they are up. He went up because he is paying a lot of attention or because he is putting all his effort, but when they go down, I get worried. Why? If last year they were doing better and now all of a sudden they are down! All that concerns me—if the children are paying attention at school, if they are having worries. They even ask me if the children have had any abuse. Or if I have problems. I tell them that in my home no one else lives here except us. The children don't have any problem with me. Sometimes I reprimand them—I even take the step to punish them, but not in a physical manner . . . . |
|---|---|
| Interviewer: | And, and, is there some ways in which you help your son learn math at home? |
| Rosa: | Yes. We start counting in many different ways. |
| Interviewer: | For example? |
| Rosa: | Like sometimes, we start playing as if we are little chickens—well, I am mother hen, and you are the little chicks. I tell them, "If they take off three little chickens from momma hen, how many chickens is she left with?" They say that many, and I say okay. "And if they add four chicks to mother hen," I say, "how many chicks will she have?" And he says, "Well, if we are four and we add another four, then eight." "That is addition," I tell them. "Okay, let's subtract," and you see that's how he enjoys those games. But he likes high numbers. He doesn't like, "Oh, Mother, I already know that." He does math fast. "How much is two times two, minus two?" He says, "I already know all of that." "Well, if you want to play high numbers, we begin to play how much is 1,000 times this minus 1,000 or 2,000 minus 3,000," and there he begins. Or we start playing with toys, and we put all the toys together, and we manage to put 100 or 200 little toys, and that's how we begin. Yes, I have this many toys, and Uriel has that many, and Emily has so many minus amongst all that. And I tell them, "Let's multiply." . . . I give him a multiplication, and he sits in the middle with all of us around him. And we begin to share, and that's it. I tell him to multiply as much as we'll all have together, and he loves it—he loves it very much. |

Look back at what you wrote in Figure 13.3 and think about the ways in which the right- and left-hand columns are similar and different, and how these differences can affect the relationship between teachers and multilingual parents/families.

As noted earlier in this chapter, sometimes there is a mismatch between teachers' and parents' expectations of one another's roles in children's education (Rodríguez-Brown, 2010). In a study about teachers' and parents' expectations of parent engagement, Scribner, Young, and Pedroza (1999) found that teachers often defined parent engagement from an academic standpoint, whereas multilingual parents defined it from a moral and social standpoint. Teachers usually think parents should engage with their children's school by "participating in activities such as school events, meetings, workshops, and governance activities, and working as teacher aides, tutors, and school advocates within the larger school community" (Scribner et al., 1999, p. 37). For multilingual parents, however,

> informal activities at home were identified as the most important parent contributions to children's success in school. Checking homework assignments, reading and listening to children read, obtaining tutorial assistance, providing nurturance, instilling cultural values, talking with children, and sending them to school well fed, clean, and rested were among the informal activities parents saw as involvement with the educational process. (Scribner et al., 1999, p. 37)

According to Vera and colleagues (2012), "parental expectations and communication about the value of school are more powerful influences than are more overt types of parental involvement (e.g., checking homework)" (p. 197).

Consider the following common misconceptions about multilingual parents' engagement:

▶ Multilingual parents are "frequently perceived as lacking resources (e.g., experience, know-how, and education) to provide and support home educational experiences for their children" (Arias & Morillo-Campbell, 2008, p. 8; Vera et al., 2012).

▶ "Many educators assume that lack of parental participation is evidence of lack of parental interest" (Arias & Morillo-Campbell, 2008, p. 8; McWayne, Manz, & Ginsburg-Block, 2015).

▶ "Traditional cultural values and beliefs of Hispanic families focus on relationships and not on competitive factors such as academic achievement. These values can contrast sharply with those of the mainstream U.S. educational system, in which individualism, self-reliance, and academic achievement are held in high regard (Perea, 2004)" (Good, Masewicz, & Vogel, 2010, p. 322). This difference could be interpreted as multilingual parents not being interested in their children's education.

## STOP AND THINK

Stop and think about when you encounter these misconceptions in your school or community.

• How do you respond?

## REFRAMING MULTILINGUAL PARENTAL ENGAGEMENT

Some authors suggest that we should move from parent involvement to parent engagement, which positions parents with agency and as active partners with schools. Looking for culturally relevant and culturally sustaining definitions of family engagement should "grant agency to families in decisions about their children's education" (Lowenhaupt, 2014, p. 523). Thus, the focus on parent engagement is identifying how multilingual families can lead at school (Dorner et al., 2019).

> *We should move from parent involvement to parent engagement, which positions parents with agency and as active partners with schools.*

### STOP AND THINK

Stop and think about what multilingual parents encounter in your school setting.

- What challenges may multilingual parents encounter when they want to engage with school activities and with their children's mathematics assignments?

## STRATEGIES FOR FACILITATING MULTILINGUAL FAMILY AND PARENT ENGAGEMENT IN YOUR CLASSROOM

Many different factors may affect the ways in which multilingual parents engage with schools or with school subjects, particularly with mathematics. Some research studies suggest that there is a disconnect between the home way and the school way of doing things. For example, some parents know different approaches to doing arithmetic than what is being taught to their children (Civil & Planas, 2010).

Prior research has identified that high-performing schools with large migrant populations have the following characteristics: School personnel visit families in the homes, and schools invest resources into parent education by "(a) increasing parental awareness of school procedures and community resources; and (b) providing self-improvement training where parents can acquire skills that may help them secure jobs outside of agricultural work" (López, Scribner, & Mahitivanichcha, 2001, p. 282). Schools viewed these investments as "human and environmental support for a child's educational development" (López et al., 2001, p. 282). In addition to these characteristics, there are other ways you can engage with multilingual parents and families in your classroom.

▶ Never assume families' needs; rather, ask parents and families what they need. Most importantly, you should never assume that a parent who does not speak English has limited schooling or is a newcomer to the United States because that might not be the case. Some families enter the United States with advanced education, such as medical

> *Never assume families' needs; rather, ask parents and families what they need.*

degrees, from their home country. Other families may have been in the United States for more than one generation. You should make every effort to develop relationships with families and not make assumptions about students' backgrounds, socioeconomic status, experiences, prior schooling, knowledge, culture, and so on. Instead, learn how immigrant households and communities function (Bonney et al., 2019; Noguerón-Liu, Hall, & Smagorinsky, 2017).

▶ Find innovative methods to share information and communicate. Modify the ways in which opportunities and resources for parent engagement at school and at home are made available to all parents (Lee & Bowen, 2006), and give space for questions when families are unfamiliar with the U.S. school system. For example, make phone calls or send printed notes in their home language when you consider it necessary. Note that in some cultures, a note sent home from the teacher indicates that something negative has occurred or that the child is in trouble at school. Therefore, it is important to explain to multilingual learners and their families that U.S. schools frequently send home written information. It is also helpful to establish communication with multilingual families early wherein messages sent from school are positive, rather than waiting until there may be a problem to communicate with the family.

▶ Find effective and efficient ways to communicate with families, since multilingual parents may feel alienated when school staff, teachers, or translators do not speak the family's primary language (Arias & Morillo-Campbell, 2008; Fantuzzo et al., 2013). For instance, many schools invite other community members to serve as translators at school events so that parents who do not speak English or are learning English feel heard and included (Bonney et al., 2019). However, it's important to advertise the availability of these services to parents who may not know they exist (Vera et al., 2012). You can also use common greetings such as *hello* or *welcome* in other languages. A quick internet search can provide you with video tutorials for virtually any language. This simple greeting may release stress and create more confidence to communicate for both you and multilingual families.

▶ Value other languages at school and encourage students to use their primary language when they find it necessary (i.e., translanguaging). For example, you can encourage students to teach the class words or phrases in their first language. The class can also learn about differences in mathematics in other cultures.

▶ Provide a welcoming environment for parents from diverse cultural and linguistic backgrounds at school. Design parent events with activities where parents and children can work together and multilingual parents are part of the organization team (Robles, 2011). Think beyond common formats, such as multicultural night, and let parents take a leadership role (Bonney et al., 2019).

▶ Recognize and respect the heritage of all families. Create opportunities for parents to engage in authentic conversations about their children to increase their empowerment and sense of equality (Good et al., 2010). For example, identify the child and family strengths and integrate these as the focus of your engagement with families. Value interaction patterns of multilingual families such as where parents and grandparents share stories through oral history (Souto-Manning & Swick, 2006).

▶ Conduct classroom visits with parents. In this approach, a small group of parents and one or two facilitators (researchers, school–community liaison, school administrator) visit a mathematics classroom and then debrief their observations, with or without the teacher (Civil & Quintos, 2009). Make sure to create spaces for parents to discuss issues related to teaching and learning mathematics, and listen to their questions and concerns (Civil & Menendez, 2010).

▶ Provide guidance on the kinds of questions parents should be asking their child to facilitate learning (e.g., "Can you draw a picture?"; Acosta-Iriqui, Diez-Palomar, Marshall, & Quintos, 2011), and explain to all families the purpose of homework. Advise families when you will move to a new topic. Also, emphasize concepts that will be revisited.

▶ Pose the question that Booker and Goldman (2016) asked families: "What will aid families in advocating for their children's math learning?" Booker and Goldman found that engaging in this discussion changed parents' agency and authority about math.

▶ Show families what their child is capable of via evidence (Acosta-Iriqui et al., 2011). For example, have students share their knowledge with families through homework. The following examples are taken from Acosta-Iriqui and colleagues (2011):

  ○   Students take home a mathematical game to play with their families. Students should write the directions for the game along with strategies for how to win the game.

  ○   Students can write mathematical problems for their families to solve at parent/family night.

## REFLECTING ON FAMILY ENGAGEMENT IN YOUR PRACTICE

Your first interactions with multilingual families are critical. In Try It! 13.3, reflect on past experiences with first interactions.

 **Try It! 13.3**

When multilingual families walk into your *school* for the first time, what should they experience? When they walk into your *classroom*, what should they experience? Think about the first five minutes of your first interaction with one multilingual family. For example, this could occur at an open house, on the first day of school, when a family moves into the community, or during parent–teacher conferences. What will you do differently?

**Reflect**

- How will you support multilingual family engagement in your mathematics classroom and school community?

# APPENDIX A
## RECOMMENDED RESOURCES

If you would like to read more about specific topics, here are some recommended resources:

## PROFESSIONAL ORGANIZATIONS AND CENTERS

Center for the Mathematics Education of Latinos/as
    https://cemela.math.arizona.edu/
Colorín Colorado—A bilingual site for educators and families of English language learners
    www.colorincolorado.org
English Learners Success Forum
    www.elsuccessforum.org
Mathematics for English Language Learners
    www.tsusmell.org
National Council of Teachers of Mathematics
    www.nctm.org
TODOS: Mathematics for ALL—Excellence and Equity in Mathematics
    www.todos-math.org

## PARTICIPATION

Schultz, K. (2009). *Rethinking classroom participation: Listening to silent voices*. Teachers College Press.

## CULTURALLY RELEVANT CONTEXTS

Gay, G. (2002). Preparing for culturally responsive teaching. *Journal of Teacher Education, 53*, 106–116.

Gutstein, E., Lipman, P., Hernandez, P., & de los Reyes, R. (1997). Culturally relevant mathematics teaching in a Mexican American context. *Journal for Research in Mathematics Education, 28*, 709–737.

Domínguez, H. (2016). Mirrors and windows into student noticing. *Teaching Children Mathematics, 22*(6), 358–365.

Ladson-Billings, G. (1995). But that's just good teaching! The case for culturally relevant pedagogy. *Theory Into Practice, 34*(3), 159–165.

Ladson-Billings, G. J. (1999). Preparing teachers for diverse student populations: A critical race theory perspective. *Review of Research in Education, 24*, 211–247.

Moschkovich, J. (1999). Supporting the participation of English language learners in mathematical discussions. *For the Learning of Mathematics, 19*(1), 11–19.

Vomvoridi-Ivanovic, E. (2012). Using culture as a resource in mathematics: The case of four Mexican-American prospective teachers in a bilingual after-school program. *Journal of Mathematics Teacher Education, 15*, 53–66.

## STUDENTS' UNDERSTANDING OF MATHEMATICS

Carpenter, T., Franke, M., & Levi, L. (2003). *Thinking mathematically: Integrating arithmetic and algebra in elementary school.* Heinemann.

Chval, K., Lannin, J., & Jones, D. (2013). *Putting essential understanding of fractions into practice in grades 3–5.* National Council of Teachers of Mathematics.

Chval, K., Lannin, J., & Jones, D. (2016). *Putting essential understanding of geometry and measurement into practice in grades 3–5.* National Council of Teachers of Mathematics.

Crites, T., Dougherty, B. J., Slovin, H., & Karp, K. (2018). *Putting essential understanding of geometry into practice in grades 6–8.* National Council of Teachers of Mathematics.

de Araujo, Z., Dougherty, B. J., & Zenigami, F. (2018). *Putting essential understanding of expressions and equations into practice in grades 6–8.* National Council of Teachers of Mathematics.

Lannin, J., Chval, K., & Jones, D. (2013). *Putting essential understanding of multiplication and division into practice in grades 3–5.* National Council of Teachers of Mathematics.

Molina, M., & Ambrose, R. C. (2006). Fostering relational thinking while negotiating the meaning of the equals sign. *Teaching Children Mathematics, 13,* 111–117.

Olson, T. A., Olson, M., & Slovin, H. (2015). *Putting essential understanding of ratios and proportions into practice in grades 6–8.* National Council of Teachers of Mathematics.

## CURRICULUM

Chval, K. B., & Chávez, Ó. (2011). Designing math lessons for English language learners. *Mathematics Teaching in the Middle School, 17,* 261–265.

Moschkovich, J. (2012, June 11). ELLs and the Common Core in mathematics. *Understanding Language.* http://ell.stanford.edu/event/ells-and-common-core-mathematics

## PARENT AND FAMILY ENGAGEMENT

Booker, A., & Goldman, S. (2016). Participatory design research as a practice for systemic repair: Doing hand-in-hand math research with families. *Cognition and Instruction, 34*(3), 222–235.

Caspe, M., Lopez, M. E., & Hanebutt, R. (2019). *The family engagement playbook.* https://medium.com/familyengagementplaybook

# APPENDIX B
## SELECTED SOLUTIONS

## SOLUTIONS FROM CHAPTER 7

Here are some answers to the number sentences from Figure 7.1.

Number Sentence 1:

5 △(6) + 2 ♡(5) = 9 ⬡(4) + 4 ◇(1)

5 △(8) + 2 ♡(0) = 9 ⬡(4) + 4 ◇(1)

5 △(4) + 2 ♡(5) = 9 ⬡(2) + 4 ◇(3)

Number Sentence 2:

3 (17) + 6 (17) + 2 (17) = 5 [35] + 12

3 (12) + 6 (12) + 2 (12) = 5 [24] + 12

3 (7) + 6 (7) + 2 (7) = 5 [13] + 12

## SOLUTIONS FROM CHAPTER 9

Meanings for expressions used in the 1960s and 2020s:

| 1960s | 2020s |
|---|---|
| It's a gas: It's a lot of fun | He's so extra: He's trying too hard; over the top |
| That's boss: That's cool or neat | Big yikes: Refers to something so embarrassing that one "yikes" won't express the larger emotion so a "big" yikes is necessary |
| Pedal pushers: Cropped pants | |
| Drop a dime: Make a phone call | |
| Going steady: In a long, serious relationship | Cap/no cap: Lie/no lie |
| Padiddle: What you say when one car's headlight is out | Throw shade: Subtle way of disrespecting someone |
| | That party was lit: That party was exciting or really good |
| | You slay me: You overwhelm me |

# References

Abedi, J., & Lord, C. (2001). The language factor in mathematics tests. *Applied Measurement in Education, 14*, 219–234.

Accurso, K., Gebhard, M., & Purington, S. B. (2017). Analyzing diverse learners' writing in mathematics: Systemic functional linguistics in secondary pre-service teacher education. *International Journal for Mathematics Teaching & Learning, 18*(1), 84–108.

Acosta-Iriqui, J. M., Diez-Palomar, J., Marshall, M. E., & Quintos, B. (2011). Conversations around mathematics education with Latino parents in two borderland communities. In K. Tellez, J. Moschkovich, & M. Civil (Eds.), *Latinos/as and mathematics education* (pp. 125–147). Information Age.

Adoniou, M., & Qing, Y. (2014). Language, mathematics and English language learners. *Australian Mathematics Teacher, 70*(3), 3–13.

Alibali, M. W., Knuth, E. J., Hattikudur, S., McNeil, N. M., & Stephens, A. C. (2007). A longitudinal examination of middle school students' understanding of the equal sign and equivalent equations. *Mathematical Thinking and Learning, 9*, 221–247.

Alibali, M. W., Nathan, M. J., & Fujimori, Y. (2011). Gestures in the mathematics classroom: What's the point? In N. L. Stein & S. W. Raudenbush (Eds.), *Developmental cognitive science goes to school* (pp. 219–234). Routledge.

Allen, R., & Chval, K. (2009). Becoming environmentally aware. *Teaching Children Mathematics, 16*(1), 28–33.

Anderson, D., Stuart, M., Abadi, M., & Gal, S. (2019, January 5). 5 everyday hand gestures that can get you in serious trouble outside the U.S. *Business Insider.* https://www.businessinsider.com/hand-gestures-offensive-different-countries-2018–6

Anhalt, C. O., & Ondrus, M. (2011). Algebraic and geometric representations of perimeter with algebra blocks: Professional development for teachers of Latino English language learners. In K. Téllez, J. Moschkovich, & M. Civil (Eds.), *Latinos/as and mathematics education: Research on learning and teaching in classrooms and communities* (pp. 195–214). Information Age.

Arias, M., & Morillo-Campbell, M. (2008). *Promoting ELL parental involvement: Challenges in contested times.* Education Policy Research Unit. http://epsl.asu.edu/epru/documents/EPSL-0801–250-EPRU.pdf

Avalos, M. A., Medina, E., & Secada, W. G. (2015). Planning for instruction: Increasing multilingual learners' access to algebraic word problems and visual graphics. In A. Bright, H. Hansen-Thomas, & L. C. de Oliveira (Eds.), *The Common Core State Standards in mathematics and English language learners: High school* (pp. 5–28). National Council of Teachers of Mathematics.

Baker, C., & Wright, W. W. (2017). *Foundations of bilingual education and bilingualism* (6th ed.). Multilingual Matters.

Bakhtin, M. (1986). *Speech genres and other late essays*. University of Texas Press.

Ball, D. L., & Forzani, F. M. (2011). Building a common core for learning to teach: And connecting professional learning to practice. *American Educator, 35*(2), 17–39.

Barnett-Clarke, C., Fisher, W., Marks, R., & Ross, S. (2010). *Developing essential understanding of rational numbers for teaching mathematics in grades 3–5*. National Council of Teachers of Mathematics.

Barwell, R. (2003). Patterns of attention in the interaction of a primary school mathematics student with English as an additional language. *Educational Studies in Mathematics, 53*(1), 35–59.

Barwell, R. (2009). Mathematical word problems and bilingual learners. In R. Barwell (Ed.), *Multilingualism in mathematics classrooms: Global perspectives* (pp. 1–13). Multilingual Matters.

Bay-Williams, J. M., & Livers, S. (2009). Supporting MATH vocabulary acquisition. *Teaching Children Mathematics, 16*, 238–245.

Behr, M., Erlwanger, S. H., & Nichols, E. (1976). *How children view equality sentences: PMDC Technical Report No. 3* (ED144802). ERIC. https://eric.ed.gov/?id=ED144802

Bonney, E. N., Dorner, L., Trigos-Carrillo, L., Song, K., & Kim, S. (2019). Developing inclusive and multilingual family literacy events at diverse schools. In E. R. Crawford & L. Dorner (Eds.), *Educational leadership of immigrants: Case studies in times of change* (pp. 176–182). Routledge.

Booker, A., & Goldman, S. (2016). Participatory design research as a practice for systemic repair: Doing hand-in-hand math research with families. *Cognition and Instruction, 34*, 222–235.

Brenner, M. E. (1998). Development of mathematical communication in algebra problem solving groups: Focus on language minority students. *Bilingual Research Journal, 22*(3–4), 149–174.

Brisk, M. E. (2011). Learning to write in the second language: K–5. In *Handbook of research in second language teaching and learning* (pp. 58–74). Routledge.

Brown, A. L., Ash, D., Rutherford, M., Nakagawa, K., Gordon, A., & Campione, J. C. (1993). Distributed expertise in the classroom. In G. Salomon (Ed.), *Distributed cognitions: Psychological and educational considerations* (pp. 188–228). Cambridge University Press.

Brozo, W. G., & Crain, S. (2018). Writing in math: A disciplinary literacy approach. *The Clearing House: A Journal of Educational Strategies, Issues and Ideas, 91*(1), 7–13.

Carpenter, T., Fennema, E., & Franke, M. (1996). Cognitively guided instruction: A knowledge base for reform in primary mathematics instruction. *The Elementary School Journal, 97*(1), 3–20.

Carpenter, T. P., Franke, M. L., & Levi, L. (2003). *Thinking mathematically: Integrating arithmetic and algebra in the elementary school.* Heinemann.

Castellón, L., Burr, L., & Kitchen, R. (2011). English language learners' conceptual understanding of fractions: An interactive interview approach as a means to learn with understanding. In K. Téllez, J. Moschkovich, & M. Civil (Eds.), *Latinos/as and mathematics education: Research on learning and teaching in classrooms and communities* (pp. 259–282). Information Age.

Castro, E., Exposito-Casas, E., Lopez-Martin, E., Lizasoain, L., Navarro-Ascencio, E., & Gaviria, J. (2015). Parental involvement on student academic achievement: A meta-analysis. *Educational Research Review, 14*(1), 33–46.

Cazden, C. B., & Beck, S. W. (2003). Classroom discourse. In A. Graesser, M. Grensbacher, & S. Goldman (Eds.), *Handbook of discourse processes* (pp. 165–197). Lawrence Erlbaum.

Celedón-Pattichis, S., & Turner, E. E. (2012). "Explícame tu respuesta": Supporting the development of mathematical discourse in emergent bilingual kindergarten students. *Bilingual Research Journal, 35*(2), 197–216.

Chval, K. B. (2004, October). Tools for thought and communication. In D. McDougall (Ed.), *Proceedings of the twenty-sixth annual meeting of the North American Chapter of the International Group of the Psychology of Mathematics Education* (Vol. 3, pp. 1473–1479). Ontario Institute for Studies in Education, University of Toronto.

Chval, K. (2010, March). *Mathematics curriculum and Latino English language learners: Moving the field forward* [Paper presentation]. CEMELA-CPTM-TODOS Conference, Tucson, AZ.

Chval, K. B. (2012). Facilitating the participation of Latino English language learners: Learning from an effective teacher. In S. Celedón-Pattichis & N. G. Ramirez (Eds.), *Beyond good teaching: Advancing mathematics education for ELLs* (pp. 77–90). National Council of Teachers of Mathematics.

Chval, K. B., & Chávez, Ó. (2011). Designing math lessons for English language learners. *Mathematics Teaching in the Middle School, 17*(5), 261–265.

Chval, K. B., Chávez, Ó., Pomerenke, S., & Reams, K. (2009). Enhancing mathematics lessons to support all students. In D. Y. White & J. S. Silva (Eds.), *Mathematics for every student: Responding to diversity PK–5* (pp. 43–52). National Council of Teachers of Mathematics.

Chval, K. B., & Hicks, S. (2012). Strategically using calculators in the elementary grades. In C. Hirsch, G. Lappan, & B. Reys (Eds.), *Curriculum issues in an era of common core state standards for mathematics* (pp. 125–137). National Council of Teachers of Mathematics.

Chval, K., & Khisty, L. L. (2001, April). *Writing in mathematics with Latino fifth-grade students* [Paper presentation]. Annual meeting of the American Educational Research Association, Seattle, WA.

Chval, K. B., & Khisty, L. L. (2009). Latino students, writing, and mathematics: A case study of successful teaching and learning. In R. Barwell (Ed.), *Multilingualism in mathematics classrooms: Global perspectives* (pp. 128–144). Multilingual Matters.

Chval, K. B., Lannin, J., & Jones, D. (2013). *Putting essential understandings of fractions into practice in grades 3–5*. National Council of Teachers of Mathematics.

Chval, K., & Pinnow, R. (2010). Preservice teachers' assumptions about Latino/a English language learners. *Journal of Teaching for Excellence and Equity in Mathematics, 2*(1), 6–12.

Chval, K. B., & Pinnow, R. J. (2018). A path to discourse-rich communities. *Teaching Children Mathematics, 25*(1), 105–112.

Chval, K. B., Pinnow, R. J., Smith, E., & Rojas Perez, O. (2018). Promoting equity, access, and success through productive student partnerships. In S. Crespo, S. Celedón-Pattichis, & M. Civil (Eds.), *Access and equity: Promoting high-quality mathematics in grades 3–5* (pp. 115–132). National Council of Teachers of Mathematics.

Chval, K. B., Pinnow, R. J., & Thomas, A. (2015). Learning how to focus on language while teaching mathematics to English language learners: A case study of Courtney. *Mathematics Education Research Journal, 27*(1), 103–127.

Chval, K. B., & Reys, R. (2008). Effective use of manipulatives across the elementary grade levels: Moving beyond isolated pockets of excellence to school-wide implementation. *Journal of Mathematics Education Leadership, 10*(1), 3–8.

Cipolla, L. (2018, September 6). 6 hand gestures in different cultures (& what they mean). *Busuu Blog*. https://blog.busuu.com/what-hand-gestures-mean-in-different-countries/

Civil, M., & Menendez, J. M. (2010). *Involving Latino and Latina parents in their children's mathematics education*. National Council of Teachers of Mathematics.

Civil, M., & Planas, N. (2010). Latino/a immigrant parents' voices in mathematics education. *Immigration, Diversity, and Education*, pp. 130–150.

Civil, M., & Quintos, B. (2009). Latina mothers' perceptions about the teaching and learning of mathematics: Implications for parental participation. *Culturally Responsive Mathematics Education*, pp. 321–343.

Cobb, P. (1995). Mathematical learning and small-group interaction: Four case studies. In P. Cobb & H. Bauersfeld (Eds.), *The emergence of mathematical meaning: Interaction in classroom cultures* (pp. 25–129). Lawrence Erlbaum.

Cohen, E. G., & Lotan, R. A. (2004). Equity in heterogeneous classrooms. In J. Banks & C. McGee Banks (Eds.), *Handbook of research on multicultural education* (2nd ed., pp. 736–750). Jossey-Bass.

Collier, V. P. (1995). *Promoting academic success for ESL students: Understanding second language acquisition for school.* New Jersey TESOL—Bilingual Educators.

Crawford, E. R., & Dorner, L. M. (Eds.). (2019). *Educational leadership of immigrants: Case studies in times of change.* Routledge.

Cummins, J. (2008). BICS and CALP: Empirical and theoretical status of the distinction. In B. Street & N. H. Hornberger (Eds.), *Encyclopedia of language and education* (2nd ed., Vol. 2, pp. 71–83). Springer Science + Business Media.

Danielsson, K. (2016). Modes and meaning in the classroom: The role of different semiotic resources to convey meaning in science classrooms. *Linguistics and Education, 35,* 88–99. doi: 10.1016/j.ling.2016.07.005

David Pakman Show. (2015, June 27). *The rapidly changing language of American English* [Video]. YouTube. https://www.youtube.com/watch?v=4YSbNaXaOy0

Davison, D., & Pearce, D. (1990). Perspectives on writing activities in the mathematics classroom. *Mathematics Education Research Journal, 2*(1), 15–22.

de Araujo, Z., Roberts, S. A., Willey, C., & Zahner, W. (2018). English learners in K–12 mathematics education: A review of the literature. *Review of Educational Research, 88,* 879–919.

de Jong, E. J., & Harper, C. A. (2005). Preparing mainstream teachers for English-language learners: Is being a good teacher good enough? *Teacher Education Quarterly, 32,* 101–124.

de Jong, E. J., Harper, C. A., & Coady, M. R. (2013). Enhanced knowledge and skills for elementary mainstream teachers of English language learners. *Theory Into Practice, 52*(2), 89–97.

Denmark, T. (1976). *Final report: A teaching experiment on equality: PMDC Technical Report No. 6* (ED144805). ERIC. https://eric.ed.gov/?id=ED144805

Dettlaff, A. J. (2008). Immigrant Latino children and families in child welfare: A framework for conducting a cultural assessment. *Journal of Public Child Welfare, 2,* 451–470.

Dillon, J. J. (1982). The multidisciplinary study of questioning. *Journal of Educational Psychology, 74,* 147–165.

Doerr, H. M., & Chandler-Olcott, K. (2009). Negotiating the literacy demands of standards-based curriculum materials. In J. Remillard, B. Herbel-Eisenmann, & G. Lloyd (Eds.), *Mathematics teachers at work: Connecting curriculum materials and classroom instruction* (pp. 283–301). Routledge.

Domínguez, H. (2005). Bilingual students' articulation and gesticulation of mathematical knowledge during problem solving. *Bilingual Research Journal, 29,* 269–293.

Domínguez, H. (2011). Using what matters to students in bilingual mathematics problems. *Educational Studies in Mathematics, 76*(3), 305–328. https://doi.org/10.1007/s10649-010-9284-z

Domínguez, H. (2016). Mirrors and windows into student noticing. *Teaching Children Mathematics, 22*, 358–365.

Domínguez, H., López Leiva, C. A., & Khisty, L. L. (2014). Relational engagement: Proportional reasoning with bilingual Latino/a students. *Educational Studies in Mathematics, 85*(1), 143–160. https://doi.org/10.1007/s10649-013-9501-7

Dorner, L., Song, K., Kim, S., & Trigos-Carrillo, L. (2019, November/December). Multilingual family engagement: Shifting the focus from what families need to how they can lead. *Literacy Today*, pp. 30–31.

Dougherty, B. J. (1996). The write way: A look at journal writing in first-year algebra. *The Mathematics Teacher, 89*, 556–560.

Enyedy, N., Rubel, L., Castellón, V., Mukhopadhyay, S., Esmonde, I., & Secada, W. (2008). Revoicing in a multilingual classroom. *Mathematical Thinking and Learning, 10*, 134–162. doi: 10.1080/10986060701854458

Fantuzzo, J., Gadsden, V., Li, F., Sproul, F., McDermott, P., Hightower, D., & Minney, A. (2013). Multiple dimensions of family engagement in early childhood education: Evidence for a short form of the Family Involvement Questionnaire. *Early Childhood Research Quarterly, 28*(4), 734–742.

15 hand gestures that have different meanings overseas. (2019). *Bright Side.* https://brightside.me/wonder-places/15-hand-gestures-that-have-different-meanings-overseas-769110/

Folse, K. S. (2004). *Vocabulary myths: Applying second language research to classroom teaching.* University of Michigan Press.

Folse, K. S. (2008). Six vocabulary activities for the English language classroom. *English Teaching Forum, 46*(3), 12–21.

Fosnot, C. T. (2007). *The T-shirt factory: Place value, addition, and subtraction.* Firsthand Heinemann.

Franke, M. L., Webb, N. M., Ing, M. M., Chan, A. G., & Freund, D. (2007, April). *Understanding student mathematical learning: Relationships among mathematical tasks, teacher practice, student participation, and student achievement* [Paper presentation]. Annual meeting of the American Educational Research Association, Chicago, IL.

The Fred Rogers Company. (1981). *How people make crayons* [Video]. *Mister Rogers' Neighborhood.* https://www.misterrogers.org/articles/factory_visits/

Freedawn Home. (2015, October 30). *Inside the Crayola factory where 12m crayons are made every day* [Video]. YouTube. Retrieved from https://www.youtube.com/watch?v=fK3-su7aQ5w

Freeman, B., Higgins, K. N., & Horney, M. (2016). How students communicate mathematical ideas: An examination of multimodal writing using digital technologies. *Contemporary Educational Technology, 7*, 281–313.

Fuson, K. C. (1998). Pedagogical, mathematical, and real-world conceptual-support nets: A model for building children's multidigit domain knowledge. *Mathematical Cognition, 4*(2), 147–186.

Garcia, E., & Gonzalez, R. (1995). Issues in systemic reform for culturally and linguistically diverse students. *Teachers College Record, 96*, 418–431.

García, O., & Lin, A. M. (2017). Translanguaging in bilingual education. *Bilingual and Multilingual Education*, pp. 117–130.

Garet, M. S., Porter, A. C., Desimone, L., Birman, B. F., & Yoon, K. S. (2001). What makes professional development effective? Results from a national sample of teachers. *American Educational Research Journal, 38*(4), 915–945.

Gebhard, M., Chen, I. A., & Britton, L. (2014). "Miss, nominalization is a nominalization": English language learners' use of SFL metalanguage and their literacy practices. *Linguistics and Education, 26*, 106–125.

Gee, J., & Gee, J. P. (2007). *Social linguistics and literacies: Ideology in discourses.* Routledge.

Gee, J. P. (1999). *An introduction to discourse analysis: Theory and method.* Routledge.

Gee, J. P. (2015). Discourse, small d, big D. In K. Tracy (Ed.), *The international encyclopedia of language and social interaction* (Vol. 1, pp. 418–422). Wiley-Blackwell.

Genesee, F. (2006). *Educating English language learners: A synthesis of research evidence.* Cambridge University Press.

Gibbons, P. (2009). *English learners, academic literacy, and thinking: Learning in the challenge zone.* Heinemann.

Gibbons, P. (2015). *Scaffolding language, scaffolding learning* (2nd ed.). Heinemann.

González, N., Moll, L. C., & Amanti, C. (Eds.). (2005). *Funds of knowledge: Theorizing practices in households, communities, and classrooms.* Lawrence Erlbaum.

Good, M. E., Masewicz, S., & Vogel, L. (2010). Latino English language learners: Bridging achievement and cultural gaps between schools and families. *Journal of Latinos and Education, 9*, 321–339.

Gottlieb, M., & Ernst-Slavit, G. (2019). *Academic language in diverse classrooms: Mathematics, grades 3–5.* Corwin.

Graham, S., McKeown, D., Kiuhara, S., & Harris, K. R. (2012). A meta-analysis of writing instruction for students in the elementary grades. *Journal of Educational Psychology, 104*(4), 879–896.

Gullberg, M. (2008). Gestures and second language acquisition. In P. Robinson & N. Ellis (Eds.), *Handbook of cognitive linguistics and second language acquisition* (pp. 276–305). Routledge.

Gullberg, M. (2010). Methodological reflections on gesture analysis in second language acquisition and bilingualism research. *Second Language Research, 26*(1), 75–102.

Gutierrez, R. (2002). Beyond essentialism: The complexity of language in teaching mathematics to Latina/o students. *American Educational Research Journal, 39*(4), 1047–1088.

Gutstein, E. (2003). Teaching and learning mathematics for social justice in an urban, Latino school. *Journal for Research in Mathematics Education, 34,* 37–73.

Gutstein, E., Lipman, P., Hernandez, P., & de los Reyes, R. (1997). Culturally relevant mathematics teaching in a Mexican American context. *Journal for Research in Mathematics Education, 28,* 709–737.

Hansen-Thomas, H. (2009). Reform-oriented mathematics in three 6th grade classes: How teachers draw in ELLs to academic discourse. *Journal of Language, Identity & Education, 8*(2–3), 88–106. doi: 10.1080/15348450902848411

Harré, R. (2012). Positioning theory: Moral dimensions of socio-cultural psychology. In J. Valsiner (Ed.), *The Oxford handbook of culture and psychology* (pp. 191–206). Oxford University Press.

Harré, R., & van Langenhove, L. (1999). The dynamics of social episodes. In R. Harré & L. van Langenhove (Eds.), *Positioning theory: Moral contexts of intentional action* (pp. 1–13). Blackwell.

Haynes, J. (2005). Challenges for ELLs in content area learning. *Everything ESL.* http://ftp.everythingesl.net/inservices/challenges_ells_content_area_l_65322.php

Helfrich, S. R., & Bean, R. M. (2011). Beginning teachers reflect on their experiences being prepared to teach literacy. *Teacher Education and Practice, 24*(2), 201–222.

Henningsen, M., & Stein, M. K. (1997). Mathematical tasks and student cognition: Classroom-based factors that support and inhibit high-level mathematical thinking and reasoning. *Journal for Research in Mathematics Education, 28,* 524–549.

Herbel-Eisenmann, B. A. (2007). From intended curriculum to written curriculum: Examining the "voice" of a mathematics textbook. *Journal for Research in Mathematics Education, 38,* 344–369.

Herbel-Eisenmann, B. A., Steele, M. D., & Cirillo, M. (2013). (Developing) teacher discourse moves: A framework for professional development. *Mathematics Teacher Educator, 1*(2), 181–196.

Herbel-Eisenmann, B., Wagner, D., Johnson, K., Suh, H., & Figueras, H. (2015). Positioning in mathematics education: Revelations on an imported theory. *Educational Studies in Mathematics, 89,* 185–204.

Hiebert, J., & Wearne, D. (1992). Links between teaching and learning place value with understanding in first grade. *Journal for Research in Mathematics Education, 23,* 98–122.

Hill, J., & Flynn, K. (2006). *Classroom instruction that works with English language learners* (pp. 14–22). Association for Supervision and Curriculum Development.

Hirsch, C. R., Lappan, G., & Reys, B. (Eds.). (2012). *Curriculum issues in an era of Common Core State Standards for mathematics*. National Council of Teachers of Mathematics.

Hirvela, A. (2011). Writing to learn in content areas: Research insights. In R. Manchón (Ed.), *Learning-to-write and writing-to-learn in an additional language* (pp. 159–180). John Benjamins.

Horn, I. (2012). *Strength in numbers: Collaborative learning in secondary mathematics*. National Council of Teachers of Mathematics.

Huang, J., Normandia, B., & Greer, S. (2005). Communicating mathematically: Comparison of knowledge structures in teacher and student discourse in a secondary math classroom. *Communication Education, 54*(1), 34–51.

Hutchins, P. (1986). *The doorbell rang*. Greenwillow Books.

Hyland, K. (2007). Genre pedagogy: Language, literacy and L2 writing instruction. *Journal of Second Language Writing, 16*(3), 148–164.

Hyland, K. (2019). *Second language writing*. Cambridge University Press.

Jago, C. (2014). Do no harm. *Voices From the Middle, 21*(3), 10–12.

Jenkins, R. (2014). *Social identity*. Routledge.

Jhagroo, J. R. (2015). I know how to add them, I didn't know I had to add them. *Australian Journal of Teacher Education, 40*(11), 107–119. doi: 10.14221/ajte.2015v40n11.6

Jukin Media. (2012, March). *Girls first ski jump* [Video]. YouTube. https://youtu.be/ebtGRvP3ILg

Kayi-Aydar, H. (2015). Teacher agency, positioning, and English language learners: Voices of pre-service classroom teachers. *Teaching and Teacher Education, 45*, 94–103.

Kendon, A. (2004). *Gesture: Visible action as utterance*. Cambridge University Press.

Kersaint, G., Thompson, D. R., & Petkova, M. (2014). *Teaching mathematics to English language learners*. Routledge.

Khisty, L. L. (1995). Making inequality: Issues of language and meanings in mathematics teaching with Hispanic students. In W. G. Secada, E. Fennema, & L.B. Adajian (Eds.), *New directions for equity in mathematics education* (pp. 279–297). Cambridge University Press.

Khisty, L. L., & Chval, K. (2002). Pedagogic discourse and equity in mathematics: When teachers' talk matters. *Mathematics Education Research Journal, 14*(3), 154–168.

Kieran, C. (1981). Concepts associated with the equality symbol. *Educational Studies in Mathematics, 12*, 317–326.

Knipper, K., & Duggan, T. (2006). Writing to learn across the curriculum: Tools for comprehension in content area classes. *The Reading Teacher, 59*, 462–470.

Knuth, E. J., Stephens, A. C., McNeil, N. M., & Alibali, M. W. (2006). Does understanding the equal sign matter? Evidence from solving equations. *Journal for Research in Mathematics Education, 37*, 297–312.

Kobett, B. M., & Karp, K. S. (2020). *Strengths-based teaching and learning in mathematics: Five teaching turnarounds for grades K–6*. Corwin.

Kosko, K. W., & Zimmerman, B. S. (2019). Emergence of argument in children's mathematical writing. *Journal of Early Childhood Literacy, 19*(1), 82–106.

Labov, W. (2011). *Principles of linguistic change: Vol. 3. Cognitive and cultural factors*. Wiley-Blackwell.

Ladson-Billings, G. J. (1999). Preparing teachers for diverse student populations: A critical race theory perspective. *Review of Research in Education, 24*, 211–247.

Ladson-Billings, G. (2014). Culturally relevant pedagogy 2.0: Aka the remix. *Harvard Educational Review, 84*(1), 74–84.

Lannin, J., Chval, K., & Jones, D. (2013). *Putting essential understandings of multiplication and division into practice in grades 3–5*. National Council of Teachers of Mathematics.

Lave, J., & Wenger, E. (1991). *Situated learning: Legitimate peripheral participation*. Cambridge University Press.

Lee, J. S., & Bowen, N. K. (2006). Parent involvement, cultural capital, and the achievement gap among elementary school children. *American Educational Research Journal, 43*, 193–218.

Lemke, J. L. (2003). Mathematics in the middle: Measure, picture, gesture, sign, and word. In M. Anderson, A. Saenz-Ludlow, S. Zellweger, & V. V. Cifarelli (Eds.), *Educational perspectives on mathematics as semiosis: From thinking to interpreting to knowing* (pp. 215–234). Legas.

Lipka, J., Sharp, N., Adams, B., & Sharp, F. (2007). Creating a third space for authentic biculturalism: Examples from math in a cultural context. *Journal of American Indian Education, 46*(3), 94–115.

Lipka, J., Sharp, N., Brenner, B., Yanez, E., & Sharp, F. (2005). The relevance of culturally based curriculum and instruction: The case of Nancy Sharp. *Journal of American Indian Education, 44*(3), 31–54.

Liu, K. K. (2015). The influence of the math classroom context on students' academic English production. *Journal of Immersion and Content-Based Language, 3*(1), 127–147. doi: 10.1075/jicb.3.1.06liu

*Long division algorithms collected in the European Union* [Paper presentation]. (2003, February/March). CERME III—Conference of the European Society for Research in Mathematics Education, Bellaria, Italy. https://sites.google.com/site/algorithmcollectionproject/long-division-algorithms-collected-in-the-european-union

López, G. R., Scribner, J. D., & Mahitivanichcha, K. (2001). Redefining parental involvement: Lessons from high-performing migrant-impacted schools. *American Educational Research Journal, 38*, 253–288. http://doi.org/10.3102/00028312038002253

Lopez, N. R. (n.d.). *Mathematical notation comparisons between U.S. and Latin American countries*. TODOS: Mathematics for ALL. https://www.csus.edu/indiv/o/oreyd/ACP.htm_files/TODOS.operation.description.pdf

Lowenhaupt, R. (2014). School access and participation: Family engagement practices in the new Latino diaspora. *Education and Urban Society*, *46*, 522–547.

Lubienski, S. T. (2000). Problem solving as a means toward mathematics for all: An exploratory look through a class lens. *Journal for Research in Mathematics Education*, *31*(4), 454–482.

Lubienski, S. T. (2002). Research, reform, and equity in U.S. mathematics education. *Mathematical Thinking and Learning*, *4*(2/3), 103–125.

Lusto, C. (2012, January). *Japanese multiplication* [Video]. YouTube. https://www.youtube.com/watch?v=85VdoNpL32k

MacDonald, R., Lord, S., & Miller, E. (2019). Doing and talking mathematics: Engaging ELLs in the academic discourse of the mathematical practices. In L. C. de Oliveira, K. M. Obenchain, R. H. Kenney, & A. W. Oliveira (Eds.), *Teaching the content areas to English language learners in secondary schools* (pp. 119–133). Springer.

Manyak, P. C. (2012). Powerful vocabulary instruction for English learners. In E. Kame'enui & J. Baumann (Eds.), *Vocabulary instruction: Research to practice* (2nd ed., pp. 280–302). Guilford Press.

Marks, G., & Mousley, J. (1990). Mathematics education and genre: Dare we make the process writing mistake again? *Language and Education*, *4*, 117–135.

McNeil, N. M., & Alibali, M. W. (2005). Knowledge change as a function of mathematics experience: All contexts are not created equal. *Journal of Cognition and Development*, *6*, 285–306.

McNeil, N. M., Grandau, L., Knuth, E. J., Alibali, M. W., Stephens, A. C., Hattikudur, S., & Krill, D. E. (2006). Middle-school students' understanding of the equal sign: The books they read can't help. *Cognition and Instruction*, *24*, 367–385.

McWayne, C. M., Manz, P. H., & Ginsburg-Block, M. D. (2015). Examination of the Family Involvement Questionnaire-Early Childhood (FIQ-EC) with low-income, Latino families of young children. *International Journal of School & Educational Psychology*, *3*, 117–134.

Mercer, N. (1995). *The guided construction of knowledge: Talk amongst teachers and learners*. Multilingual Matters.

Miller, J., & Warren, E. (2014). Exploring ESL students' understanding of mathematics in the early years: Factors that make a difference. *Mathematics Education Research Journal*, *26*, 791–810. doi:10.1007/s13394-014-0121-z

Molina, M., & Ambrose, R. C. (2006). Fostering relational thinking while negotiating the meaning of the equals sign. *Teaching Children Mathematics*, *13*(2), 111–117.

Moore-Harris, B. (2005, July 7–9). *Strategies for teaching mathematics to English language learners* [Paper presentation]. International Math Conference, San Antonio, TX.

Morales, H., Khisty, L. L., & Chval, K. B. (2003). Beyond discourse: A multimodal perspective of learning mathematics in a multilingual context.

In N. Pateman, B. Dougherty, & J. Zilliox (Eds.), *Proceedings of the 2003 joint meeting of PME and PMENA* (Vol. 3, pp. 133–140). Center for Research and Development Group, University of Hawaii.

Moschkovich, J. N. (1999). Supporting the participation of English language learners in mathematical discussions. *For the Learning of Mathematics, 19*(1), 11–19.

Moschkovich, J. (2002). A situated and sociocultural perspective on bilingual mathematics learners. *Mathematical Thinking and Learning, 4,* 189–212.

Moschkovich, J. (2012, January). Mathematics, the Common Core, and language: Recommendations for mathematics instruction for ELLs aligned with the Common Core [Paper presentation]. *Understanding language: Commissioned papers on language and literacy issues in the Common Core State Standards and Next Generation Science Standards* (pp. 17–31). Understanding Language Conference, Stanford University, Stanford, CA.

Moschkovich, J. (2013). Principles and guidelines for equitable mathematics teaching practices and materials for English language learners. *Journal of Urban Mathematics Education, 6*(1), 45–57.

Moschkovich, J. N. (2015). Academic literacy in mathematics for English learners. *The Journal of Mathematical Behavior.* doi: http://dx.doi.org/10.1016/j.jmathb.2015.01.005

National Academies of Sciences, Engineering, and Medicine. (2018a). *English learners in STEM subjects: Transforming classrooms, schools, and lives.* The National Academies Press. doi: https://doi.org/10.17226/25182

National Academies of Sciences, Engineering, and Medicine. (2018b). *How people learn II: Learners, contexts, and cultures.* The National Academies Press. doi: https://doi.org/10.17226/24783

National Council of Teachers of Mathematics. (2000). *Principles and standards for school mathematics.* https://www.nctm.org/Standards-and-Positions/Principles-and-Standards/

National Council of Teachers of Mathematics. (2014). *Principles to actions: Ensuring mathematical success for all.* https://www.nctm.org/PtA/

NBC Universal. (2017, March 9). *Inside the Crayola factory: See how the iconic crayons are made* [Video]. *Today.* https://www.today.com/video/inside-the-crayola-factory-see-how-the-iconic-crayons-are-made-893853251852

Noguerón-Liu, S., Hall, D., & Smagorinsky, P. (2017). Building on immigrant parents' repertoires. In S. Salas & P. Portes (Eds.), *U.S. Latinization: Education and the new Latino South* (pp. 1–22). SUNY Press.

Ogbu, J. U., & Simons, H. D. (1998). Voluntary and involuntary minorities: A cultural-ecological theory of school performance with some implications for education. *Anthropology and Education Quarterly, 29*(2), 155–188.

O'Neill, D. K., Topolovec, J., & Stern-Cavalcante, W. (2002). Feeling sponginess: The importance of descriptive gestures in 2- and 3-year-old children's acquisition of adjectives. *Journal of Cognition and Development, 3*(3), 243–277.

Orellana, M. F. (2016). *Immigrant children in transcultural spaces: Language, learning, and love.* Routledge.

Orey, D. C. (1999). *Welcome to the Algorithm Collection Project.* http://www.csus.edu/indiv/o/oreyd/acp.htm_files/alg.html

Page, D. (1964). *Number lines, functions, and fundamental topics: Mathematics for elementary school teachers.* Macmillan.

Page, D., Wagreich, P., & Chval, K. (1993). *Maneuvers with triangles.* Dale Seymour. http://www.lsritlrl.uic.edu/area-of-right-triangles-maneuvers-with-triangles-chapter-1/

Page, D., Wagreich, P., & Chval, K. (1995). *Maneuvers with circles.* Dale Seymour. http://www.lsritlrl.uic.edu/whats-inside-maneuvers-with-circles-chapter-2/

Panadero, E., & Jonsson, A. (2013). The use of scoring rubrics for formative assessment purposes revisited: A review. *Educational Research Review, 9*, 129–144.

Pappamihiel, N. (2002). English as a second language students and English language anxiety: Issues in the mainstream classroom. *Research in the Teaching of English, 36*, 327–355.

Paris, D. (2012). Culturally sustaining pedagogy: A needed change in stance, terminology, and practice. *Educational Researcher, 41*(3), 93–97.

Perea, S. (2004). *The New America: The America of the moo shoo burrito.* HIS Ministries.

Peregoy, S. F., & Boyle, O. F. (2016). *Reading, writing, and learning in ESL: A resource book for K–8 teachers* (7th ed.). Longman.

Pinnow, R., & Chval, K. (2014). Positioning ELLs to develop academic, communicative, and social competencies in mathematics. In M. Civil & E. Turner (Eds.), *Common Core State Standards in mathematics for English language learners: Grades K–8* (pp. 21–34). TESOL International Association.

Pinnow, R. J., & Chval, K. B. (2015). "How much you wanna bet?": Examining the role of positioning in the development of L2 learner interactional competencies in the mathematics classroom. *Linguistics & Education, 30*, 1–11. doi: 10.1016/j.linged.2015.03.04

Pitvorec, K., Willey, C., & Khisty, L. L. (2011). Toward a framework of principles for ensuring effective mathematics instruction for bilingual learners through curricula. In B. Atwah, M. Graven, W. Secada, & P. Valero (Eds.), *Mapping equity and quality in mathematics education* (pp. 407–422). Springer.

Raborn, D. T. (1995). Mathematics for students with learning disabilities from language-minority backgrounds: Recommendations for teaching. *New York State Association for Bilingual Education Journal, 10*, 25–33.

Radford, L. (2009). Why do gestures matter? Sensuous cognition and the palpability of mathematical meanings. *Educational Studies in Mathematics, 70*(2), 111–126.

Razfar, A., Khisty, L. L., & Chval, K. B. (2011). Re-mediating second language acquisition: A sociocultural perspective for language development. *Mind, Culture, and Activity, 18*(3), 195–215.

Robles, S. (2011). *Parental involvement in an urban minority school district* [Unpublished doctoral dissertation]. Seton Hall University.

Rodríguez-Brown, F. V. (2010). Latino families: Culture and schooling. In E. Murillo, S. Villenas, R. Galvan, J. Munoz, C. Martinez, & M. Machado-Casas (Eds.), *Handbook of Latinos and education: Theory, research, and practice* (pp. 350–360). Routledge.

Rojas-Drummond, S., Maine, F., Alarcón, M., Trigo, A. L., Barrera, M. J., Mazón, N., Vélez, M., & Hofmann, R. (2017). Dialogic literacy: Talking, reading and writing among primary school children. *Learning, Culture and Social Interaction, 12*, 45–62.

Rothstein, A., & Rothstein, E. (2007). Writing and mathematics: An exponential combination. *Principal Leadership, 7*(5), 21–25.

Rowling, J. K. (2017a). *Harry Potter and the philosopher's stone.* Bloomsbury.

Rowling, J. K. (2017b). *Harry Potter and the sorcerer's stone.* Bloomsbury.

Rymes, B. (2014). *Communicating beyond language: Everyday encounters with diversity.* Routledge.

Rymes, B. (2016). *Classroom discourse analysis: A tool for critical reflection* (2nd ed.). Routledge.

Salehmohamed, A., & Rowland, T. (2014). Whole-class interactions and code-switching in secondary mathematics teaching in Mauritius. *Mathematics Education Research Journal, 31*, 431–446.

Schleppegrell, M. J. (2004). *The language of schooling: A functional linguistics perspective.* Lawrence Erlbaum.

Scribner, J. D., Young, M. D., & Pedroza, A. (1999). Building collaborative relationships with parents. In P. Reyes, J. D. Scribner, & A. P. Scribner (Eds.), *Lessons from high-performing Hispanic schools: Creating learning communities* (pp. 36–60). Teachers College Press.

Secada, W. G., & De La Cruz, Y. (1996). Teaching mathematics for understanding to bilingual students. In J. LeBlanc Flores (Ed.), *Children of* la frontera: *Binational efforts to serve Mexican migrant and immigrant children* (pp. 285–308). ERIC Clearinghouse on Rural Education & Small Schools.

Setati, M., & Adler, J. (2000). Between languages and discourses: Language practices in primary multilingual mathematics classrooms in South Africa. *Educational Studies in Mathematics, 43*(3), 243–269. doi: 10.1023/A:1011996002062

Shein, P. P. (2012). Seeing with two eyes: A teacher's use of gestures in questioning and revoicing to engage English language learners in the repair of mathematical errors. *Journal for Research in Mathematics Education, 43*, 182–222. https://doi.org/10.5951/jresematheduc.43.2.0182

Smith, E. (2018). *Facilitating emergent bilinguals' participation in mathematics: An examination of a teacher's positioning acts* [Unpublished doctoral dissertation]. University of Missouri.

Solano-Flores, G. (2011). Language issues in mathematics and the assessment of English language learners. In K. Téllez, J. Moschkovich, & M. Civil (Eds.), *Latinos/as and mathematics education: Research on learning and teaching in classrooms and communities* (pp. 283–314). Information Age.

Souto-Manning, M., & Swick, K. J. (2006). Teachers' beliefs about parent and family involvement: Rethinking our family involvement paradigm. *Early Childhood Education Journal, 34*, 187–193.

Steffe, L., & Cobb, P. (1988). *Construction of arithmetical meanings and strategies*. Springer-Verlag.

Tabors, P. (2008). *One child, two languages: A guide for early childhood educators of children learning English as a second language* (2nd ed.). Paul H. Brookes.

Takeuchi, M. (2015). The situated multiliteracies approach to classroom participation: English language learners' participation in classroom mathematics practices. *Journal of Language, Identity & Education, 14*, 159–178. doi: 10.1080/15348458.2 015.1041341

Tavares, N. J. (2015). How strategic use of L1 in an L2-medium mathematics classroom facilitates L2 interaction and comprehension. *International Journal of Bilingual Education and Bilingualism, 18*, 319–335.

Teale, W. H. (2009). Students learning English and their literacy instruction in urban schools. *The Reading Teacher, 62*(8), 699–703.

Turner, E. E., & Celedón-Pattichis, S. (2011). Mathematical problem solving among Latina/o kindergartners: An analysis of opportunities to learn. *Journal of Latinos and Education, 10*(2), 146–169.

Turner, E. E., Dominguez, H., Empson, S., & Maldonado, L. A. (2013). Latino/a bilinguals and their teachers developing a shared communicative space. *Educational Studies in Mathematics, 84*, 349–370. doi: 10.1007/s10649-013-9486-2.

Turner, E., Dominguez, H., Maldonado, L., & Empson, S. (2013). English learners' participation in mathematical discussion: Shifting positionings and dynamic identities. *Journal for Research in Mathematics Education, 44*, 199–234.

Valdés, G. (1996). Con respeto: *Bridging the distances between culturally diverse families and schools*. Teachers College Press.

Valdés, G. (2004). Between support and marginalisation: The development of academic language in linguistic minority children. *International Journal of Bilingual Education and Bilingualism, 7*, 102–132.

Velasco, P., & García, O. (2014). Translanguaging and the writing of bilingual learners. *Bilingual Research Journal, 37*(1), 6–23.

Vera, E. M., Israel, M. S., Coyle, L., Cross, J., Knight-Lynn, L., Moallem, I., & Goldberger, N. (2012). Exploring the educational involvement of parents of English learners. *School Community Journal, 22*, 183–202.

Vomvoridi-Ivanovic, E. (2012). Using culture as a resource in mathematics: The case of four Mexican-American prospective teachers in a bilingual after-school program. *Journal of Mathematics Teacher Education, 15*, 53–66.

Vomvoridi-Ivanovic, E., & Chval, K. B. (2014). Challenging beliefs and developing knowledge in relation to teaching English language learners: Examples from mathematics teacher education. In B. Cruz, C. Ellerbrock, A. Vasquez, & E. Howes (Eds.), *Talking diversity with teachers and teacher educators: Exercises and critical conversations across the curriculum* (pp. 115–130). Teachers College Press.

Vygotsky, L. (1978). *Mind in society.* Harvard University Press.

Wager, A. (2012). Incorporating out-of-school mathematics: From cultural context to embedded practice. *Journal of Mathematics Teacher Education, 15*, 9–23.

Warren, E., & Young, J. (2008). Oral language, representations and mathematical understanding: Indigenous Australian students. *Australian Journal of Indigenous Education, 37*, 130–137.

*Washington Post.* (2014, July 9). *How do you pronounce "water"?* [Video]. YouTube. https://www.youtube.com/watch?v=Q7ijTGd6hy0

Webb, N. M., Franke, M. L., De, T., Chan, A. G., Freund, D., Shein, P., & Melkonian, D. K. (2009). "Explain to your partner": Teachers' instructional practices and students' dialogue in small groups. *Cambridge Journal of Education, 39*, 49–70.

Webb, N. M., Franke, M. L., Ing, M., Chan, A., De, T., Freund, D., & Battey, D. (2008). The role of teacher instructional practices in student collaboration. *Contemporary Educational Psychology, 33*, 360–381.

Webel, C. (2010). Shifting mathematical authority from teacher to community. *The Mathematics Teacher, 104*, 315–318.

WIDA Consortium. (2015, September). *Focus on family engagement.* https://wida.wisc.edu/sites/default/files/resource/FocusOn-Family-Engagement.pdf

Wikipedia. (2020, June 11). *Jai alai.* https://en.wikipedia.org/wiki/Jai_alai

Wilde, S. (1991). Learning to write about mathematics. *The Arithmetic Teacher, 38*(6), 38–43.

Wilder, S. (2014). Effects of parental involvement on academic achievement: A meta-synthesis. *Educational Review, 66*(3), 377–397.

Wilson, K., & Devereux, L. (2014). Scaffolding theory: High challenge, high support in academic language and learning (ALL) contexts. *Journal of Academic Language and Learning, 8*(3), A91–A100.

Wood, T. (1998). Funneling or focusing? Alternative patterns of communication in mathematics class. In H. Steinbring, M. G. Bartolini-Bussi, & A. Sierpinska (Eds.), *Language and communication in the mathematics classroom* (pp. 167–178). National Council of Teachers of Mathematics.

Yang, D. C. (2005). Developing number sense through mathematical diary writing. *Australian Primary Mathematics Classroom, 10*(4), 9–14.

Yoon, B. (2007). Offering or limiting opportunities: Teachers' roles and approaches to English-language learners' participation in literacy activities. *The Reading Teacher, 61*(3), 216–225.

Yoon, B. (2008). Uninvited guests: The influence of teachers' roles and pedagogies on the positioning of English language learners in the regular classroom. *American Educational Research Journal, 45*(2), 495–522. http://dx.doi.org/10.3102/0002831208316200

Yoon, B. (2012). Junsuk and Junhyuck: Adolescent immigrants' educational journey to success and identity negotiation. *American Educational Research Journal, 49*(5), 971–1002.

Yosso, T. J. (2005). Whose culture has capital? A critical race theory discussion of community cultural wealth. *Race Ethnicity and Education, 8*(1), 69–91.

Zahner, W. (2012). "Nobody can sit there": Two perspectives on how mathematics problems in context mediate group problem solving discussions. *REDIMAT: Journal of Research in Mathematics Education, 1*, 105–135. doi: 10.4471/redimat.2012.07

Zangori, L., & Pinnow, R. J. (2020). Positioning participation in the NGSS era: What counts as success? *Journal of Research in Science Teaching, 57*(4), 623–648.

Zarate, M. E. (2007). *Understanding parental involvement in education: Perceptions, expectations and recommendations.* Tomás Rivera Policy Institute.

Zheng, S., & Dai, W. (2012). Studies and suggestions on prewriting activities. *Higher Education Studies, 2*(1), 79–87.

Zwiers, J., & Hamerla, S. (2018). *The K-3 guide to academic conversations: Practices, scaffolds, and activities.* Corwin.

Zwiers, J., & Soto, I. (2017). *Academic language mastery: Conversational discourse in context.* Corwin.

# Index

A SAGE Publishing Company

Helping educators make the greatest impact

**CORWIN HAS ONE MISSION:** to enhance education through intentional professional learning.

We build long-term relationships with our authors, educators, clients, and associations who partner with us to develop and continuously improve the best evidence-based practices that establish and support lifelong learning.

NATIONAL COUNCIL OF
TEACHERS OF MATHEMATICS

The National Council of Teachers of Mathematics supports and advocates for the highest-quality mathematics teaching and learning for each and every student.